Specifying Buildings

A design management perspective

Specifying Buildings

A design management perspective

Second edition

Stephen Emmitt

and

David T. Yeomans

AMSTERDAM • BOSTON • HEIDELBERG • LONDON • NEW YORK
OXFORD • PARIS • SAN DIEGO • SAN FRANCISCO
SINGAPORE • SYDNEY • TOKYO
Butterworth-Heinemann is an imprint of Elsevier

Butterworth-Heinemann is an imprint of Elsevier
Linacre House, Jordan Hill, Oxford OX2 8DP, UK
30 Corporate Drive, Suite 400, Burlington, MA 01803, USA

First published 2001
Second Edition 2008

British Library Cataloguing in Publication Data
A catalogue record for this book is available from the British Library

Library of Congress Cataloging-in-Publication Data
A catalog record for this book is available from the Library of Congress

ISBN: 978-0-7506-8450-7

For information on all Butterworth-Heinemann publications
visit our web site at books.elsevier.com

Typeset by Charon Tec Ltd (A Macmillan Company)
www.charontec.com

Printed and bound in Hungary
08 09 10   10 9 8 7 6 5 4 3 2 1

Contents

1 Specifications in context

The glamorous and the dramatic aspects of building that have been dealt with in novels and turned into films concern the relationship between the architect and the client, or the building process. The first is seen in *The Fountainhead* (Rand, 1943), which is concerned with the creative process of design and seen as the glamorous aspect of building. Truer to life is *Mr Blandings Builds His Dreamhouse* (Hodgins, 1946). The film that was made of the book includes a memorable scene in which the foreman carpenter asks Mrs Blandings whether she wants the cills rebated or unrebated. Her decision to have them unrebated leads to a cascade of falling timber as the rebated cills are ripped out. She explains to her husband, 'I thought it sounded cheaper'. Here, the book and the film get to the heart of building, being precise about what it is that you want. Kidder (1985) in a more recent novel, *House*, tells the story from the point of view of the contractor, where some of the drama rises directly from poor communication between the parties. H. B. Creswell's two books *The Honeywood File* and *The Honeywood Settlement* (Creswell, 1929, 1930) deal exclusively with communication. These books, which are the imaginary letter files of an architect building a house for a rich client, although written over three quarters of a century ago, are still excellent reading for anyone involved in this process and, although presented in a lighthearted manner, take the issues that have been occasionally dealt with by novelists and present them as matters for serious consideration by professionals. Just as the routine of police work is never shown in police dramas, behind the drama of building is this routine but vital work of communicating precisely what it is that is wanted.

Building, whether it be a house or a multi million pound complex, first requires a clear specification of the client's requirements that will be translated by the architect into a design. This means effectively interpreting a client's requirements into a set of briefing documents, from which the conceptual and detailed design is developed. Design intent is then codified in the contract documentation and subsequently translated into a building. Exploring, exploiting and enhancing client requirements in the form of a creative design that satisfies planning legislation and building codes; is buildable within a set budget, safely and within a defined period; satisfies user demands and also recognizes environmental constraints is a familiar process to the design team, regardless of building size and complexity. At the heart of this process is the selection of materials, components and products that make up a building's assembly: they contribute to the aesthetics, quality and durability of the completed building. As long ago as 1933 the architect Chermayeff, in an article entitled 'New materials and new methods' (Chermayeff, 1933), emphasized the importance of specifying the correct material for a given situation when he said that:

> It is essential to select for a specific purpose within the defined cost, the most adequate material and method; that is to say, that material which best solves the problems of purpose, money and time.

This still holds true today. The specification process is 'an inherent part of the design process' and that 'any lack of thoroughness in specifying products' may cause problems during construction and create programme delays (CPIC, 2003). Analysis and choice of building products or, more specifically, the selection of the correct building products for a specific purpose, within the limits of cost and time, is an important task for the specifier because it helps to determine the quality of our built environment. Since the 1930s there have been many changes in how buildings are designed and constructed, reflecting changes in architectural fashion, developments in construction technology and more sophisticated approaches to the management of design. Although some of our buildings may look very similar to those built in the past, the reality is that buildings are now designed and built to higher performance standards and have to comply with far more extensive legislation than ever before. With the desire to build faster, cheaper and to higher standards there has been a shift in emphasis from craftsmanship applied at the building site to off site production in carefully controlled factory environments. Of course this trend is by no means universal and, currently, it is possible to find highly industrialized process existing alongside craft based activities and also to find new products and materials being used in conjunction with those that have been around for centuries. There is also a growing concern for the environmental impact of construction activities and the building in use, together with the health and well being of building users, which has helped to re emphasize the importance of selecting the best materials and methods for a healthy and sustainable architecture. Thus, the specification of buildings is a fundamental aspect of architectural design, dominating the events from the early conceptual designs to the physical realization of the building.

Designers and engineers specify their intentions, and these (along with other contractual information) are interpreted by the contractor and translated into a physical building. Drawings, models, schedules and bills of quantities cannot convey the whole message; they have to be supplemented with descriptive information. On very small projects, this information is often provided in the form of notes on drawings, but for the majority of projects, the descriptive information is extensive and needs to be in separate documentation known as a 'specification'.

Specifications are written documents that describe the requirements to which the service or product has to conform, i.e. its defined quality. It is usual practice to use the term 'specification' in the singular, which is a little misleading. In practice, the work to be carried out will be described in specifications written by the different specialists involved in the project. Even on the most simple building schemes, the engineer will write the specification for the structural elements such as foundations, the architect will be concerned with materials and finishes, and there will be an electrical and mechanical specification, possibly a specification for the landscaping, and one for highways work. This collection of multi authored information is known as 'the specification'.

The eventual quality of a building is determined by a combination of factors: the design, selection and specification of building materials and products; the accuracy of contractual information; and the ability of the contractor and subcontractors to interpret information that comes to them, whether in the form of lines, words or figures, into a completed building. It is the written specification, not the drawings, in which the quality standards to be achieved are set out. Thus, if we are to achieve and improve the quality of the buildings around us, a good starting point is to understand and manage the specification process in its entirety.

A historical note

Salzman (1952) looked at early building contracts where the client communicated directly with the builder, who was both designer and contractor, and so it might be today in simple remodelling work, usually at a domestic scale. But with the growth of professional services, architects (and surveyors) provided the designs and supervised construction. When work was in the hands of the tradespeople the guild system regulated the practice of the trades and ensured standards of workmanship, but as this system broke down ensuring quality also became the task of the professionals (Yeomans, 1988a). At first the changes in the nature of construction would not have been beyond the ability of the apprentice trained craftsmen to cope with and the specification of work might be fairly simple. Tradespeople knew what was wanted because that was the proper way to build, but gradually clients and their architects wanted buildings of increasing complexity, both in form and in the details of construction, and this required increasing attention on the part of the architect to the specification of the work. This was a process that took place gradually. In the letterbook of the eighteenth century architect William Chambers (B.L. Add. MS 41133-6) we find him giving precise instructions for the framing of a roof for the house at

Milton Abbey, while he considers that for the stables the carpenter should know himself how to build 'such a trifle'.

With time it was inevitable that the process of specification should become more complex, if only because by the turn of the nineteenth century buildings were already incorporating manufactured components. The cheaper transport of goods by canal meant that it was possible to bring cast iron components to rural building sites: structural as well as decorative elements. Even where components and materials were still worked on site the substitution of 'imported' materials meant that tradespeople might be working with materials that were no longer familiar to them, whether these materials were imported from abroad or brought from some other part of the country. Consider simple lime plaster. The nature of the raw material from which it is made will have an effect on its hydraulic properties and therefore the way in which it is worked. Tradespeople would be used to the properties of their local material, but if presented with a lime brought in from elsewhere might find it more difficult to work. The shifting nature of both the nature of construction and the supply of materials and components meant that architects and surveyors were having to pay greater attention to the specification of both workmanship and materials.

The historical aspect of this is dealt with by Davis, who sees the degree of specificity within contracts as part of what he calls 'the culture of building', in his book of that name (Davis, 2000). He shows how the degree of specificity in both written contracts and their accompanying drawings has increased with time, partly as a function of the appearance of professional designers, but also because of increasing subdivision within the building process. There has been an increasing number of both specialist trades and manufactured items.

The appearance of the annual publication *Specification* in Britain around the turn of the twentieth century is symptomatic of the growing complexity of this aspect of building procurement. There had already been something of a shift towards manufactured products, rather than those made on site or in builders' own workshops, and the twentieth century was to see a growing number of trades and types of products being used in building. The development of steel and then reinforced concrete frames, the introduction of electrical services and the need for mechanical ventilation are just some of the most significant changes in the technology of construction that necessitated increases in the topics covered by this publication. The result of this is that the scope of the documentation produced to specify the quality of building products and the standards of workmanship has become more extensive than in the past. Nevertheless, its purpose in laying down standards to be met remains the same. Likewise, books providing guidance to students and practitioners have changed little in their main message. What does tend to change is the fluctuating fashion for the use of performance over prescriptive methods and vice versa. Both approaches to specification are considered in this book, the choice of one method over another being a matter for individual design organizations and their specifiers. At the time of writing this book, performance specifications are coming back into fashion. However, adopting a performance approach does not eliminate the task of selecting proprietary products: it merely passes the decision making process down the line to the contractor and/or the sub contractors who have to make their selection from a range limited by the designer's performance parameters. Thus, the decision making process observed and described later in the book is appropriate to both prescriptive and performance specifications.

Many of the developments in specification have occurred in the past sixty years or so. In the USA, the Construction Specifications Institute (CSI) was founded in 1948 to serve the interests of specifiers and manufacturers. In 1963, the CSI and Construction Specifications Canada (CSC) worked together to implement standards and published the sixteen division Masterlist of specification sections; now the 'Masterformat'. The CSI *Manual of Practice* was first published in 1967 and revised and updated on a regular basis. Australian specifiers have seen the development and refinement of the National Specification (NATSPEC) system.

In the UK the early guides to specification were organized by craft, a good example being Donaldson's *Handbook of Specifications*, which was published in 1860. However, no specification writing standards existed until the publication in 1987 of the Common Arrangement of Works Sections (CAWS): until this time, most specifications had been arranged under the same headings as the bills of quantities (Cox, 1994). In 1987 the Co ordinating Committee for Project Information (CCPI) published a *Code of Procedure for Production Drawings* and a *Code for Procedure for Project Specifications* which set out guidance for specifiers. This was based on data collected from construction projects, which found that the biggest cause of quality problems was inadequate project information (BEDC, 1987). In this publication it was acknowledged that specification practices had to improve and that the UK did not compare too favourably with other countries, especially North America. Six years later, in the 2003 publication by the CPIC *Production Information: A Code of Procedure for the construction industry* the authors note that the development and uptake of both the National Building Specification (NBS) and the CPI project specification code had been influential in a widespread improvement in specification practice on large and medium sized projects. On a less positive note, they found that suitable specification systems for building services and small projects were not available or had not been taken up, suggesting further room for improvement.

Recently, several new pressures have been put on the specifier. One of the points noted by Latham (1994) in his much cited report into efficiency in the UK construction sector was a concern about over specification, i.e. the specification of components of a higher standard than necessary. Both the Latham report and the later Egan reports (Egan, 1998, 2002) were concerned with reducing costs and improving efficiency. Emphasis on lean thinking, the elimination of waste and improved value delivery has helped to emphasize the importance of materials specification. The promotion of partnering and collaborative, rather than competitive, forms of relationships, together with increasing use of off site manufacturing, also has implications for the way in which buildings are specified. Parallel to these changes there is a growing interest, evidenced by improved guidance (e.g. Andersen et al., 2002; Spriegel and Meadows, 2006), in the selection of green building materials and products, which implies a further change in specification practice. While this might have led to a reappraisal of the components being used and assembled and a switch to new products, there has also been an increase in risk management techniques by specifiers' offices in an attempt to reduce potential claims for negligent selection of building products. This may have resulted in the tendency to limit the number of changes away from previously used products and a reluctance to use products that are perceived as new to the office and/or specifier.

Specification research

For such an important aspect of the designer and engineer's job, it is a little surprising to find that there is very little published work in the field. Textbooks have restricted themselves to guidance on the act of writing the specification, with little or no recognition of the selection process that is an integral part of this. Much of the research that has been carried out in this area is not in the public domain. That is because the specification of building products is of great importance to the manufacturers and suppliers, who carry out a good deal of commercial research into the adoption of their own, and their direct competitors' products. Naturally, the results are commercially sensitive, which helps to explain why published research is rare.

Mackinder (1980) undertook some significant research into how architects specified building products. Information about detailed design decisions was collected from diaries filled in by participating architects and supported with interviews. She observed that architects frequently used 'short cuts' based on their own experience to save time, reporting a strong preference for certain materials and components that they had used previously, drawn from their personal collections of literature. This supported the earlier observations of Goodey and Matthew (1971) and Wade (1977). When asked about this the architects surveyed claimed that such behaviour was necessary because they had limited time in which to make decisions. Therefore it was easier and quicker to use products that were known to perform and, more importantly, not to fail. Specifiers perceived reliance on familiar manufacturers and products as a method of keeping their exposure to risk to a minimum. One third of Mackinder's sample acknowledged that new materials and methods needed to be monitored, but claimed that it was office policy to avoid the use of anything new unless it was unavoidable, preferring to specify familiar products. Proprietary and performance specifications were used concurrently. Mackinder's study found that apart from work relating to information flow and retrieval, plus a small amount of work on specification writing, there had been very little research into how professionals actually selected building products. The situation has changed little since 1980. Apart from a few small reports (Walton Markham Associates, 1981; Moore, 1987) and work by Emmitt (1997, 2001, 2006), which forms the background to this book, this crucial area continues to be overlooked by the academic research community.

The Barbour Index, a commercial supplier of information to the constructor sector, has been conducting and publishing research into specification decisions on an annual basis since 1993. The Barbour Reports cover a variety of different aspects of specification; the underlying theme is concerned with communicating information about products between manufacturer and specifier (Barbour Index, 1993, 1994, 1995, 1996, 1997, 1998, 1999, 2000, 2001, 2002, 2003, 2004, 2006). Asking specifiers how they behave via questionnaire surveys and interviews with a wide range of specifiers forms the basis of the Reports. While providing a very useful source of information for building product manufacturers keen to communicate effectively with specifiers, the data also help to reveal trends in specification practice. The data support the earlier work of Mackinder; that the majority of specifiers (regardless of professional background) act in a conservative manner, preferring to use products and manufacturers with which they are familiar. The data also show that architects and engineers remain the most important specifier, but that their influence is declining as main and specialist contractors take on more responsibility for product decisions (Barbour Index, 2006). This shift in responsibility appears to have changed quite rapidly between the 2000 and 2004 Barbour Reports as changes to procurement methods (promoted via the Latham and Egan Reports) take effect.

Learning to specify

A natural question to ask is: to what extent do students learn the art of specification during their studies? The answer to this question varies depending on the chosen programme of study. Mackinder (1980) looked at the extent to which schools of architecture taught the selection of building products and found that they did not. Schools teach what they think of as design, although it could be argued that material and product selection is an important aspect of this, but it is not much considered. This has been criticized by, for example, Crosbie (1995) and Antoniades (1992), who believed the 'major weakness' of architectural design in schools of architecture to be the lack of attention given to the importance of materials and building technology. The effect is that the schools teach what has been defined as phase 1 (conceptual design), leaving phases 2 and 3 to be learned in practice (see Figure 1.1). This appears to be a common model in both Europe and the USA, one exception being Australia, where the Royal Australian Institute of Architects (RAIA) is explicit in the requirement for members to demonstrate competency in specifications (Gelder, 1995, 2001). The habits of product selection are therefore passed on during a young designer's 'apprenticeship' within the design office and so will be strongly influenced by practice procedures and the influence of more experienced members of the organization. Thus, the tendency for specifiers to select the same products used by their colleagues and their office is strong. Without any discussion on specification practice during education, architects will be ill equipped to take a detached view of such procedures; this may well result in the perpetuation of obsolete behaviour that has become nothing more than a ritual. Abe and Starr (2003) have articulated similar concerns in engineering programmes, noting the isolation of the specification exercise from the act of designing.

The recent introduction of undergraduate degrees in Architectural Technology during the 1990s in the UK should help in this regard. These degrees incorporate specification skills as part of the architectural technologist's competency and, combined with education in design management techniques, should enable graduates be well equipped to challenge existing practices (Emmitt, 2002). According to *The Architectural Technology Careers Handbook* (CIAT, 2006) the architectural technologist is responsible for 'producing, analysing and advising upon specification, materials selection and detailed design solutions in relation to performance and production criteria', while the architectural technician's role is to 'prepare specifications for construction work'.

Learning to specify is one of the first tasks undertaken in the professional office, and young professionals tend to pick up their specification habits early in their careers. Parallels can be seen in the prescription of drugs by medical practitioners, where prescribing habits are known to form in early clinical practice. Medical schools worldwide are starting to adopt a problem based approach to learning, so that medical students can develop the skills required to evaluate critically new drugs that come onto the market (MacLeod, 1999). To encourage this approach, the World Health Organization has produced a teaching aid, *Guide to Good Prescribing* (WHO, 1995), which is designed to help students to develop a method for selecting appropriate drugs and be less susceptible to external influences, such as pressure from drug companies. A similar argument could be made for students of building design and construction, all of whom need to explore and develop knowledge of specification practices so that they are able to deal with external influences in a professional manner. This knowledge will need to be implemented and tested in practice, updated and further developed through education and training programmes as part of a planned professional development programme.

To operate efficiently, specifiers require rapid access to current and relevant information; they must possess adequate technical knowledge, have the ability to convey instructions clearly and unambiguously to the builder, and (subject to the appropriate contractual arrangement) must ensure that products are not substituted or performance standards compromised without their knowledge and approval. As demonstrated in later chapters of this book, the specification process is complex, difficult to observe in its entirety and demanding in terms of its management. These are issues taken up later; first, we need to look at the framework in which the specifier has to operate.

A framework for specifying

The manner in which a building is designed and then built is rarely a neat and ordered process; there are frequent changes, redesigns and reprogramming as the project moves from the initial idea to a finished building. Several conceptual project plans exist that aim to guide the designer through a project, the most widely known of which is the RIBA's 'Plan of Work', which implicitly divides the design process into eleven separate stages. Although the plan of work has been criticized because it does not allow for feedback loops, which would allow new information to be incorporated in the ongoing decision sequences, it continues to be used by practitioners as a guide to organization and resourcing of projects, as well as a guide for fee invoicing. Research has found that designers do not adhere rigidly to the plan, but work closely to it (e.g. Mackinder, 1980). From the perspective of the specifier, there is a number of distinct phases in the specification process during which the individual will be engaged in activities that differ from the preceding stage. These may be conveniently divided into three distinct phases, namely conceptual design, detail design and production (Figure 1.1), although it is recognized that in practice there may be some overlap between these. American readers will note that the American Institute of Architects' Handbook (AIA, 1988) has five stages in an architectural project: schematic design, design development, construction documents, bidding or negotiation, and construction. For simplicity, the latter three stages have been grouped under production.

Phase 1: Conceptual design

It is at the briefing stage where performance parameters for the intended building should be agreed and confirmed in writing to form the design brief.

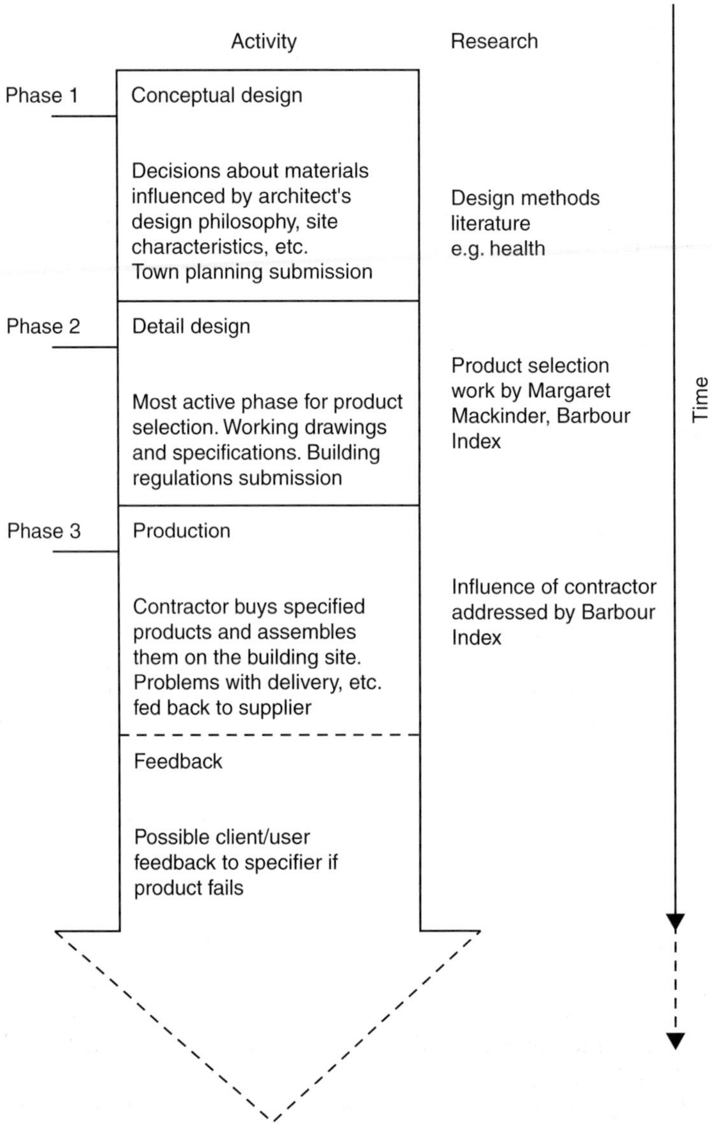

Fig. 1.1 *Product selection and the design process*

The designer should also be considering the life cycle of the building, its maintenance and final disposal strategy at this early stage, but there is little evidence to show how this is seriously considered in design except when issues of serviceability and durability are involved. Sustainability is a relatively new idea, an innovation as a design concept, and while there may be considerable lip service paid to the idea, it is questionable whether it has had much influence on actual design processes (Emmitt, 2000). Following the production of a working brief, a feasibility study and sketch designs are produced (plans and elevations) for client approval before submission for planning consent. Here, it is common to confirm generic materials to be used on the external faces of the building (e.g. 'red facing brickwork'; 'profiled metal cladding, steel blue') and, therefore, some material selection is occurring early in the process, coincident with the architect's overall design vision and the need to obtain consents from the client and the town planners. The point is that these initial decisions about major materials will influence the decisions made in phase 2.

Phase 2: Detail design

This involves the production of a large number of drawings and schedules that will enable the production of the bills of quantities, enable selected contractors to negotiate or bid a contract price and enable the chosen contractor to build the building. During the production of the detail drawings (comprising assembly drawings, component drawings and schedules), the designer will be making decisions that relate to specific building product selection. Detailed input from consultants such as structural engineers, quantity surveyors and also the client may influence specification decisions during this phase. Furthermore, it is during this phase that conformity with building codes, regulations and standards must be ensured. At least, that is the conventional picture in the UK because it is in this phase that there may be different organizational arrangements. For example, there is an increasing use of performance specifications by architects, which leaves the choice of actual product to the contractor and sub contractors. In France, the activities involved in these first two phases will actually be carried out by different kinds of office, the architects who produce the design and the *bureau d'études*, who produce the production drawings. Indeed, this separation of responsibilities between offices can also be found within larger offices elsewhere, where the detailing and specification writing is carried out by someone other than the designer of the overall concept.

Although numerous decisions are being made throughout the design process that will inform the writing of the specification, the most active stage of any project in terms of specification decision making is the detail design phase. This phase involves the production of a large number of drawings and schedules, which will both enable the production of the bills of quantities and be used by the contractor to construct the building. When the project enters the detailed design stage, the decision making process is different from that in the conceptual stage for a number of reasons. First, the designer will be constrained to a certain extent by the conceptual design and any performance parameters that have already been established; thus, the problems may be better defined than in the earlier stages. Secondly, a number of individuals will be involved in the development of the details – other consultants, manufacturers, contractors, cost consultants and project managers – and this will influence decisions that lead to the writing of the specification.

During the detailed design phase, the designer will be trying to finalize the construction details, so the relationship between detailing, product selection and the writing of the specification is very close. Because the resulting information will be used by the contractor to assemble the building, any errors or discrepancies between the documents may well lead to disputes, litigation and arbitration.

It is during phase 2 that specifiers will be actively looking for solutions to their particular detailing problems and therefore may be more receptive to information about building products that are new to them. At the end of phase 2, the specification must be complete, so decisions about how to detail particular junctions, and hence what products and/or performance levels are required, must be determined and confirmed in the written specification. At the tendering stage and throughout the construction phase, there may be pressure to change materials and products (for a variety of reasons), and so the decision-making process may well extend beyond phase 2 and into phase 3.

Phase 3: Production

This is essentially a phase concerned with the realization of the design in the form of a finished product, the translation of abstract ideas to a physical artefact, the building. Although all specification decisions should have been confirmed by the end of phase 2, there may be pressure to change the level of performance or specified products to cheaper ones, possibly because the cost of the building is over budget. But pressure to change may also come from the contractor and sub contractors because of supply difficulties, to improve buildability on site and/or to reduce product costs. It is also possible for changes to be imposed by client bodies and external control agencies. The way in which this occurs has been illustrated by Andrews and Taylor (1982), who give an account of how major changes affecting the appearance of the building may be imposed upon the architect by external factors quite beyond his or her control. More recently, a television series and accompanying book (Sabbagh, 1989) examined the construction of a New York skyscraper, which provided an insight into how the working of designers continues into the construction phase, but such 'natural' histories of the design process are regrettably few.

Consequences

Every specification decision will have consequences for the quality and durability of the building. There are also consequences for the organization that specified the products, hopefully good in terms of a job well done and repeat business, but unfortunately, sometimes, the consequences are less welcome with some form of legal action being brought against the design office. However, because of the consequence of failure, specifiers need to be aware of a wide range of issues that influence their decision making process, as do the design managers who programme and manage the process. Specifiers must have the ability to make informed decisions and communicate those decisions effectively and efficiently. The design office must have appropriate managerial frameworks and procedures in place to help specifiers to achieve their objectives quickly and accurately.

As the construction sector has become increasingly litigious, the written specification has taken on a more important role than it had in the past. Like the drawings, the written specification is a legal document and will be examined thoroughly should a dispute arise. It is necessary, therefore, for the specifier to have a thorough knowledge of contractual issues. By this, we mean an awareness of procurement options, conditions of contract, drawings, bills of quantities, standards and codes, and an ability to coordinate them all in a logical and thorough manner. In practice, this can be more challenging than it might at first appear simply because the specification writer may not have been involved in the preparation of this documentation.

Terms, definitions and scope of the book

The word specification is used in two different ways in the building sector. On the one hand, it is a term used to describe the selection of a building product by a specifier, and on the other hand, it refers to a physical document containing written descriptions of standards of workmanship and the performance of materials and products. It is not uncommon for designers and contractors to use an abbreviated version of the word specification in everyday conversation and refer to 'the spec'. To avoid confusion, *specification* is used in this book to describe the process of selection of building products and *written specification* when referring to the document. The term *specifier* is used to cover all those involved in product selection, regardless of their background.

Scope of the book

Building product selection is an essential and familiar process to practitioners, yet, in spite of the frequency with which products are specified and the importance of the resulting decision, the act of specification has rarely been observed. It is implicit in design training but is rarely taught as a distinct skill, and furthermore (and perhaps surprisingly), research into the field is sparse. Thus, there is little in the way of evidence based research on which the practitioner can draw. The aim in writing this book is to provide a generic approach to the specification of buildings so that the contents are of value to specifiers wherever they happen to practise. In taking such an approach, the authors have tried to resist the 'this is how you do it' approach; instead, research findings are used to illustrate some of the points raised in the book, from which the readers can draw their own conclusions. Because of the generic nature of the material in this book, the authors have deliberately avoided descriptions of terminology in contract documentation, contractual issues and legal matters, other than matters of general concern. Examples of standard layouts have been avoided, partly because of the generic approach and partly because information technology based software packages render such examples largely redundant. Issues relating to responsibility (and hence liability) are addressed in a generic manner, and readers are urged to seek legal advice for their particular circumstances and contract peculiarities, simply because these vary widely and will change over time as legal precedents are set and subsequently tested.

The scope of this work is relatively straightforward. The physical act of specification writing has been combined with the act of product selection and the management of the specification process. In particular, the book considers how new products are taken up by an industry, where specifiers are anxious to avoid unnecessary exposure to risk, and describes the mechanisms used by design organizations to control the information coming to them through a series of formally and informally established 'gates'. Case studies based on the observed behaviour of specifiers in architectural offices are used to illustrate issues relevant to design and construction professionals engaged in, and affected by, the specification of buildings. This book is addressed to students of architecture and built environment programmes, although it may also be of interest to young practitioners and principals/managers of design offices who need to ensure the specification process is managed professionally. The contents should also be of interest to building product manufacturers, keen to understand better the specifier's milieu.

Although the book has been designed to be read from cover to cover, many readers will want to dip in at various points to suit their particular circumstances, so a small amount of guidance may be useful. The first part of the book, Chapters 1–6, deals with the theoretical and practical issues relating to specification. Chapter 2 describes how the design team specifies design intent. The information sources available to the specifier are explained in Chapter 3, before the specification process and decision making criteria are explored in Chapter 4. The task of writing good quality specifications forms the content of Chapter 5. Chapter 6 looks at the management of the specification process from the perspective of the specifier's office. From Chapter 7 onwards there is a change in emphasis to how specifiers behave in practice, with particular emphasis on the uptake of building products that are perceived as new to the recipient. Chapter 8 investigates the relationship between manufacturers and specifiers. The way in which specifiers behave is described in detail in Chapters 9 and 10. The final chapter attempts to bring together the findings of actual behaviour with the theoretical and practical considerations.

2 Specifying design intent

An essential requirement of the professional design organization is to be able to produce clear, concise and accurate information that can be used to assemble the diverse range of materials and components into a building that meets the client's requirements and expectations. With the exception of artisans and the designer/craftsperson, designers work and communicate indirectly (Potter, 1989). Their creative work is expressed in the form of instructions to manufacturers, other consultants, contractors and sub contractors, through drawings and written documents, collectively known as 'production information'. Instructions must be clear, concise, complete, free of errors, meaningful, relevant and timely to those receiving them (which is taken up in more detail in Chapter 5). This requires an understanding of the needs of those receiving the information and the context in which it is used, be it a windswept site or the controlled environment of a production factory. Issues that might be clear in the minds of the design team are not always easy to convey in drawings and written documentation. The specifier must be careful about the level of knowledge assumed by those reading the specification.

Each building project is unique in situation, more often than not unique in design, and frequently unique in the individual members that make up the temporary project team responsible for the building's design and assembly. Before looking at quality levels for materials and workmanship, it is necessary to agree on the quality level of the finished building, i.e. the performance requirements must be established. The projected life of the building and its use are primary considerations here and should be determined (as far as possible) at the briefing stage. So, too, should the maintenance strategy, life cycle costing and disposal strategy for when the building has exceeded its design life. Different parts of the building may have different periods to replacement. In a shopping complex, for example, the basic structure and services layout will probably survive several changes in the fitting out of the shops. Obvious influences on the quality of the completed building are the client's budget and time frame, the composition of the design team and the choice of contractor. More subtle influences concern the way in which the individuals party to a building project communicate, the quality of the project information and the way in which the entire process is managed.

It is during the briefing phase when values and needs are explored and confirmed in the written brief. The written brief (essentially a specification of client needs) sets the agenda for the conceptual and detail design phases. The briefing documents often contain a mixture of performance and prescriptive information, which is challenged and often revised as the design work starts to reveal various options and solutions to problems, which may often be found to be poorly defined. Briefing is a complex activity and one that lies outside the scope of this book; however, it is important to make the point that a well managed briefing process is central to the efficient development of the design and hence the specification of buildings (Blyth & Worthington, 2001; Emmitt, 2007).

The type of procurement route to be used will also be discussed at an early stage. This will establish responsibilities for specification decisions and will also have a major influence on the specification method to be used. As the building industry has continued to specialize, and hence fragment, clients and their professional advisers are now faced with a bewildering choice of procurement routes that continue to evolve. The choice of one method over another is dependent on the size and type of project as well as on the opinions of those involved. For example, architects may favour traditional methods because they retain a fair degree of control over the design and hence the quality of the building. In such arrangements the designer and specifier may be considered to be one and the same; at least both within the same organization, if not the same person. Conversely, they may favour 'design and build' contracts because it reduces their need to become involved with the details. Similarly, contractors may favour contractor led procurement options because it gives them control over design decisions. In terms of responsibility for specifying and specification writing, the contractual routes can be divided into traditional (designer led) and non-traditional routes (contractor led, management led) in which the designer is merely a sub contractor.

Of all the information provided and used in the building process, it is the written specification that describes the quality of the final artefact. Designers specify the position, quantity and quality of the building work, they do not tell the builder 'how' to construct it; that is the contractor's responsibility (hence the need for method statements). Choice of procurement route is an important consideration in this regard because it will set the contractual obligations of the designers and the contractors with regard to site supervision. With traditional forms of procurement, there are contractual obligations for the project administrator to inspect the work. With contractor led procurement routes, such as design and build, the designers may have very little control over the quality of construction because the contractor is able to adjust specifications and substitute products without reference to the designers. For these reasons, the decision to specify using proprietary and/or performance methods cannot be separated from the type of contract used.

In practice, the determination of quality is a rather complex issue because there is a range of characteristics that must be met, and it is usually necessary to place these in some order of importance for a particular project. This will be discussed in more detail in Chapter 4. In brief, however, it should be noted at this stage that there is often a conflict of interest between the designer/specifier and the contractor, who influence the quality of the finished product at different stages of the process. Designers can define the quality of materials they require through their choice of proprietary products or through the use of performance parameters, incorporating these into the written specification. At this stage quality is dealt with, to a lesser or greater extent, through the use of the master specification combined with the use of approved and prohibited lists of products. The specifier's palette of favourite products could also be viewed as a mechanism for ensuring quality.

In contrast, the contractor is initially concerned with getting the contract, thus submitting the cheapest price to get the job, and then maximizing profits during the course of the contract, often through the substitution of products initially specified for those that generate a greater degree of profit. In this, a contractor might also have 'favourite products', being those that he can obtain at discounted prices through regular suppliers or those that he is familiar with using and so requiring less management effort on his part.

Coordinated project information

Information is central to both the design and construction process. Drawings and written documents are used to describe and define a construction project and specifiers need to understand the relationship between the written specification and the drawings, i.e. what goes where. At the heart of this body of information lies the written specification, a document central to determining the quality of the finished building. Yet this document is one of the most poorly considered and misunderstood documents. Writing the specification (and the process of specification) should not be seen as something that has to be done at the end of the detail design phase. Instead, its relationship to other documents must be recognized so that the specification can be developed during the whole design process as part of a comprehensively managed and coordinated set of information.

At its best, this project information will be clear and concise and easily understood, so that the building will be completed on time and within the budget. At its worst, poorly conceived and shoddy project information will lead to confusion, inefficiency, delay, revised work, additional expense, disputes and claims. It is a sad fact that very few projects are perfect: many are flawed by poorly expressed requirements. With the advent of computers and digital information, one could be forgiven for thinking that inadequate information would become a thing of the past, but many within the industry have noted a significant increase in the quantity of information provided and a steady decline in its quality (e.g. CPIC, 2003). Unfortunately, computers make it easy to transfer errors from one document to another very quickly. But it would be misleading to blame the technology. The biggest enemy of those trying to produce comprehensive, good-quality information is that precious commodity, time. With increased pressures on time (and fees) has come the requirement to compress the amount of time taken to produce the project information; in these circumstances mistakes and omissions are to be expected.

The major bodies in the British construction sector established the Co-ordinating Committee for Project Information (CCPI) in an attempt to improve the quality of project information, and hence reduce uncertainty and disruption to construction activities. The CCPI has been instrumental in producing guidance for the design team, developing the Common Arrangement of Work Sections for Building Works (CAWS) and contributing to the Standard Method of Measurement (SMM7). Central to the ethos of the CCPI is compatibility between the drawings, the written specifications and the bills of quantities, i.e. each artefact should complement the others when read as a set of information.

Coordinated project information (CPI) is a system that categorizes drawings and written information (specifications) and is used in British Standards and in the measurement of building works, the SMM7. This relates directly to the classification system used in the National Building Specification (NBS). One of the conventions of CPI is CAWS, which has superseded the traditional sub-division of work by trade sections. The CAWS system lists around 300 different classes of work according to the operatives who will do the work; indeed, it was designed to assist the dissemination of information to sub contractors. This allows bills of quantities to be arranged according to CAWS. Items coded on drawings, in schedules and in bills of quantities can be annotated with reference back to the specification. Under CPI it is very clear that the specification is the central document in the information chain (Fig. 2.1).

Drawings

Drawings are the most familiar medium and are regarded as one of the most effective ways of communicating information. Production drawings ('blueprints') are the main vehicle of communicating the physical layout of the design and the juxtaposition of components to those responsible for putting it all together on site. Referred to as contract information or production information, this set of drawings is often complex and extensive. Not only does it take a great deal of time and skill to produce the drawings and coordinate them with those produced by other consultants, it is also a skill to read all the information contained

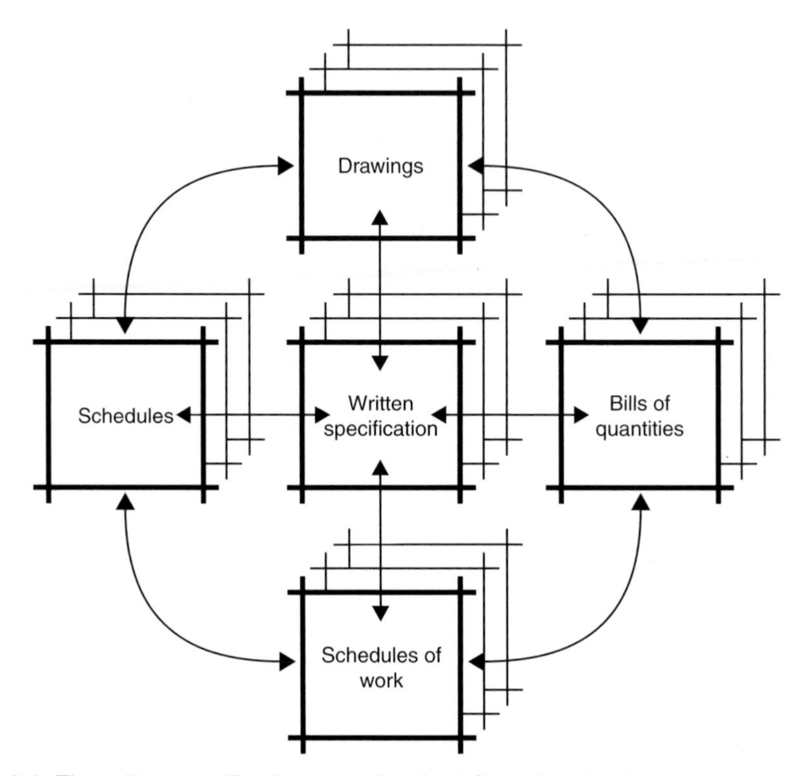

Fig. 2.1 *The written specification: central to the information circuit*

and encoded in lines, figures and symbols. It is this set of drawings that the main contractor will use to cost the building work and (subject to any revisions before starting work) this will be the set of drawings from which the building will be assembled. At its most basic, the contract drawings will comprise drawings produced by the architect, the structural engineer and the mechanical and electrical consultants. In very large building projects other contributors to this set of drawings may include interior designers, landscape designers, specialist sub contractors, highways consultant, etc.

Coordination of drawings with other consultants' information and the specification is an important consideration. A drawing system that aims to reduce repetition and overcome defects in less well coordinated systems is the *elemental method.* This is based on a four category system, starting with the location drawings, focusing on the assembly drawings, then the component drawings and finally the schedules. Each drawing has a code and a number relating to the CI/SfB construction classification system, which originated in Sweden, and is widely used in the UK. There are four codes, namely L for location, A for assembly, C for component and S for schedule. This system allows specific reference to drawings and schedules to be easily incorporated in the specification, thus aiding coordination. Another aid to coordination is the use of consistent terminology, clear cross referencing and avoidance of repetition.

The Electronic Product Information Co operation (EPIC) is also used for construction projects in Europe. So too is the Uniclass system, which is based on the principles set out in the Internationl Organization for Standardization (ISO) for the classification of construction information.

Specification notes on drawings

At this juncture, it is necessary to comment on the practice of writing notes on drawings. On very small projects and alteration works to existing buildings, it is common practice to write specification notes on the drawings, using a standard written specification to cover only the typical clauses common to most projects. Although widely used as a means of conveying information to the contractor, it is not good practice because the drawing very quickly becomes overloaded with information, repetition is largely unavoidable, and the majority of the notes are rarely descriptive enough to cover all the information required. There is a real danger that those reading the drawing on the site will rely entirely on the (incomplete) notes on the drawing and will not refer to the written specification, as they should. Apart from the obvious dangers of ineffective communication between designer and builder, this means that the drawing must be revised and reissued every time there is a change to the specification, no matter how minor. It is considered best practice to keep notes on drawings to an absolute minimum and keep the written description of materials (size, colour, manufacturer, etc.) and workmanship firmly where they belong, in the written specification.

Written documents

Written documents have always taken precedent over drawings. Until relatively recently, it was common practice to award contracts on little more than a written description of what was required (indeed, this is still common in domestic repair and alteration work where the client directly employs an organization to do work on their property, e.g. replacing the windows). The advantage of written documents, theoretically at least, is that people can understand them easier than they can drawings. Of course, this assumes that the document is well written and easy to read. Computers and word processing software have made the task of preparing and transmitting written material relatively easy; unfortunately, it is just as easy to proliferate errors.

The designer needs to be clear about the distinction between information that is best conveyed in a drawing (or drawings) and that which needs to be included in a written document. One of the authors carried out a small project making structural modifications to the house of a lawyer who decided that he could manage the project himself. As part of this he attempted to write a contract that included the information contained in the drawings. He explained that this was because he understood words; he certainly could not read drawings. Even after it had been pointed out to him that the required screws were at 25 mm spacing and not 25 mm diameter (which he then realized was rather large), and he had been introduced to the simple phrase 'supply and fix', his text was still incomprehensible without the drawings and added no more information than they contained.

Schedules

Schedules are a useful tool when describing locations in buildings where there is a repetition of information that would be too cumbersome to put on drawings. Particularly well suited to computer software spreadsheets, a schedule is a written document that lists the position of repetitive elements, such as structural columns, windows, doors, drainage inspection chambers and room finishes. For example, rooms are given their individual code and listed on a finishes schedule that will relate room number and use with the required finish of the ceiling, walls and floor.

Schedules of work

It is common practice in repair and alteration works to use a schedule of works. This document describes a list of work items to be done. It is a list that the contractor can also use for costing the work. While it is common practice to append the schedule of works to the specification, it must not be confused with the specification (see below) or, for that matter, schedules (as described above).

Bills of quantities

The bills of quantities are derived from the drawings, schedules and specification. Usually compiled by the quantity surveyor (QS) or cost consultant, their purpose is to present information in a format that is easy for the contractor's estimator to price. Bills of quantities are used on medium sized to large projects. It is unusual to produce bills of quantities for small projects because the information is usually concise, and the estimator can price the project from the drawings, schedules and specification. Although computer software packages are available to generate the bills of quantities from the designer's information, it is still common practice for a third party to prepare them, and sometimes this is the QS. In doing so, the QS frequently finds discrepancies and omissions within the information provided, and he or she thus provides a useful (unpaid) cross checking service for the design team. The contractor's estimator also has a duty to point out any deficiencies in the documentation to the design team.

The written specifications

Drawings indicate the quantity of materials to be used and show their finished relationship to each other. It is the written specification that describes the quality of the workmanship, the materials to be used and the manner in which they are to be assembled, and quality control should be foremost in the mind of the designer. This is an issue in itself and lies outside the scope of this book, but it is important to note that while the required quality can be specified in the contract documents, one needs to consider how it is going to be achieved on site and, in some cases, policed.

Like drawings, specifications vary in their size, layout and complexity. In all but the smallest of design offices, it is common for specifications to be written by someone other than the designer; thus, communication between designer and specification writer is particularly important. The majority of designers are visually orientated people whose skills are best employed in the conceptual and detailed design phases. Therefore, few have the time or the inclination to be involved in the physical writing of the document: this task is usually undertaken by a technologist or specification writer, someone with better technical and managerial skills than many designers possess.

Specification writers require an appreciation of the designer's intention and the ability to write technical documents clearly, concisely and accurately. They also need to be able to cross reference items without repetition. Standard formats form a useful template for specifiers and help to ensure a degree of consistency as well as saving time. In the UK, the NBS is widely used because it helps to save time in this way and is familiar to other parties to the design and assembly process (see Chapter 5). There is a danger in using standard specification clauses in that the specifier might simply not understand what he or she is asking for. A curious case was reported by a timber merchant who said that he had been asked to supply timber that was 'free from wane', which was reasonable enough, but also 'free from knots, shakes, sapwood and heartwood'. The first two of these were possible, but would be expensive, while the last two left only bark, which he did not deal in.

Specification methods

There are several methods available to the specifier for specifying design intent. Some methods allow the contractor some latitude for choice and therefore an element of competition in the tendering process, while others are deliberately restrictive. The terminology used tends to vary and can be confusing; therefore, this book uses the terms set out in the Construction Specifications Institute's *Manual of Practice* (CSI, 1996), which describes four methods.

- *Descriptive specifying*, where exact properties of materials and methods of installation are described in detail. Proprietary names are not used, hence this method is not restrictive.
- *Reference standard specifying*, where reference is made to established standards to which processes and products must comply, for example a national or international standard. This is also non restrictive.
- *Proprietary specifying*, where manufacturers' brand names are stated in the written specification. Here the contractor is restricted to using the specified product unless the specification is written in such a way to allow substitution of an equivalent. Proprietary specification is the most popular method, where the designer produces the design requirements and specifies in detail the materials to be used (listing proprietary products), methods and standard of workmanship required.
- *Performance specifying*, where the required outcomes are specified together with the criteria by which the chosen solution will be evaluated. This is non-restrictive and the contractor is free to use any product that meets the specified performance criteria. Performance specification is where the designer describes the material and workmanship attributes required, leaving the decision about specific products and standards of workmanship to the contractor.

The specifier's task is to select the most appropriate method which both suits a particular situation and the project context. The type of funding arrangement for the project and client preferences usually influences this decision. Typically, projects funded with public funds will have to allow for competition and so proprietary specifying is not usually possible. Projects funded from private sources may have no restrictions, unless the client has a preference or policy of using a particular approach. The client's requirements need to be considered alongside the method best suited to describe clearly the design intent and the required quality; while also considering which method will help to achieve the best price for the work and, if desired, allow for innovation. In some respects this also concerns the level of detail required for a project or particular elements of that project. Another consideration is the implied liability that a particular method presents to the specifier's office.

Although one method is usually dominant for a project, it is not uncommon to use a mix of methods for different items in the same document. Specifiers should, however, be careful to avoid redundancy (and confusion) by resisting the temptation to use a mix of methods for the same item (see Chapter 5). It would be misleading to suggest that one method of specification is better than another; instead, different situations require different approaches. The advantages and disadvantages of using one over the other are discussed below.

Descriptive specifications

The descriptive method allows the specifier to specify exactly what is required, without using proprietary names. For example, if the paint colour and finish required to the internal walls is magnolia, matt finish, then that is what is described. The method does require some expertise in the use of words, or else the intentions can become rather ambiguous. It is not uncommon for the specifier also to make reference to standards to avoid confusion. For example, magnolia paint can vary slightly in shade between different manufacturers, and therefore reference to a standard may be necessary. Descriptive specifications tend to be time consuming to write and their use can result in a bulky written specification. However, they may be very useful for small projects, especially work to existing buildings.

Such specifications are particularly important in conservation work where it is important that repairs be carried out in a manner and using materials that match, or at least are compatible with, the original. In specifying the materials to be used it is clearly the designer's responsibility to determine the nature of the original material and to describe this or the compatible material to the contractor. In some cases this will involve laboratory testing to determine its composition or properties. Examples here include the specification of mortars that will have similar porosity and water absorption characteristics to those in the original wall, or the moisture content of timbers to avoid excessive movement after they have been fixed. This is clearly a specialist area in both design and execution of the work, and the degree to which the designer needs to specify either materials or standards of workmanship will depend on the experience and knowledge of the contractor; specialist contractors are easier to specify for than general contractors with less experience of this kind of work.

Descriptions of standards of workmanship are perhaps more difficult. Phrases such as 'to match existing' can be used, but it is important that the contractor knows what this means and it is a sensible precaution to require samples to be made so that this can be tested. A difficulty here is that methods of workmanship used with present day materials are not necessarily appropriate for materials that are now only used in conservation work. For example, pointing with lime mortar requires a different technique from that used with cement based mortars. Thus, the specifier may need to consider whether the method as well as the final result needs to be described. However, it is equally important that the specifier knows what standards of workmanship actually are. One of the authors came across an incident where the architect had specified an adzed finish, expecting to see something that showed the toolmarks. He was disappointed to find the finished timbers as smooth as if they had been planed.

Reference standard specifications

Reference standard specifications are sometimes called compliance specifications, since the contractor should comply with the specified standard. Trade associations, institutions and governments publish standards to bring about consistent standards (see also Chapter 3). Standards reduce the number of types, sizes and qualities of materials as well as helping to standardize methods of testing. Some standards also provide guidance on the quality of workmanship. Specifying by reference to a standard can shorten the length of the written specification because the number of words required to describe a material and the method of testing it are considerably reduced, although some care is required in their use. Standards tend to be written for commonly used materials and products, and do not cover all situations, thus an appropriate standard may not always be available. Reference standards are a result of consensus, and in trying to include the requirements of a large number of interested parties they tend to be written to the lowest common denominator. Specifiers may require a higher performance than that stated in the appropriate standard, in which case making reference to it is meaningless. Standards that carry some authority should be used, not obscure standards that few people know. All standards contain a variety of information and options and it is not possible simply to cite a standard without first having read it to establish the relevance of the wording to the task in hand. Standards must be cited properly; that means the inclusion of the edition date of the standard, the division and the section. Simply making reference to the title or number of a standard is bad specification practice.

If there are problems with the contract that eventually result in a legal dispute it is highly likely that the arbitrator or courts will look at the published standards, even if they are not cited in the written specification. This is one reason why (too?) many specifiers splatter the written specification with references to standards to 'cover their backs'. Again, this is not good practice, especially if the specifier is not familiar with the detailed contents of the standard.

Proprietary specifications

A proprietary specification describes a product by the manufacturer's brand name, often supplemented with the manufacturer's product number and other proprietary information. This is also known as a prescriptive specification, since the product has been prescribed by the specifier for the contractor to purchase and use, just as a doctor may prescribe a brand of medicine for a patient to use. The advantage of using a proprietary specification is that the specifier has been precise in expressing his or her requirements and has control over the choice of product. For example, a facing brick might be specified as 'Ibstock, Red Rustic'. This gives the name of the manufacturer (Ibstock) and the name of the brick (Red Rustic) from the manufacturer's extensive range. It automatically gives the performance of the brick in terms of size, colour, texture, durability, water absorption, frost resistance, etc., as defined in the manufacturer's technical description. This method of specification is usually quicker for the specifier than using the performance method and is favoured by designers for materials that will be visible when the building is completed. In the example above, the specifier has been precise and knows exactly what to expect from the selection. Another advantage for the design team is that all contractors will be bidding against the same proprietary products, which should remove (in a closed proprietary specification) or reduce (in an open proprietary specification) product pricing as a major variable in the tenders.

Proprietary specifying is seen as being uncompetitive, possibly not providing value for money, and the spectre of corruption is a difficult one to shake off. One of its inherent problems is that of preventing competition. For public works, the practice is not to use proprietary specifications, the principle being that all relevant manufacturers must be able to compete for the work should they wish to do so. This is the case in both the USA and Europe. Some large organizations follow similar principles by adding the words 'or equal approved' to allow some choice by contractors and competition between suppliers.

Performance specifying and compulsory competitive tendering are seen as a way around the problem. However, proprietary specifications are widely used, and for very good reasons. For example, the recent trend towards the use of supply chain management techniques and partnering agreements necessitates a close working arrangement with chosen suppliers. Although performance standards may be used, this arrangement naturally leads to the specification of proprietary products from a particular manufacturer in the chain. Over time, the designers and manufacturers can work together to improve standards and reduce costs, but such arrangements make it difficult for others to compete. Nevertheless, the occasional use of specifying specific proprietary products means that this form of specification needs to be considered.

There have been occasions when designers have worked effectively with manufacturers to develop building components. In such circumstances the manufacturers have invested development costs and in return will expect to receive adequate orders. A good example is provided by the development of sanitary ware for Hertfordshire schools during the post war school building programme. To produce something more suitable for primary schools than the rather institutional products available on the market, the architects worked with the sanitary ware manufacturers, Adamsez, in what Saint (1987: 83) called a 'Ruskinian collaboration'. Attempts to develop school furniture in a similar way were less successful and had to wait for the development of a British Standard on school furniture.

Manufacturers spend a great deal of time and money in developing new products and/or improving products for which they usually hold patents. This does not stop their competitors from launching very similar products that are cheaper because they have incurred less research and development costs. Whether these alternative products are cheaper because they are of inferior quality is a moot point, but one worth bearing in mind, because there may be considerable pressure to change the specified product during the tendering and building phase of the contract. Furthermore, simply because a specifier has gone to a lot of trouble carefully to select and specify a proprietary product, it does not necessarily follow that the prescribed product will then be used on site. Specification substitution is a major cause for concern, both for specifiers and manufacturers (see also Chapter 6).

It is common to see product substitution as a problem, and in many cases this is exactly the case. However, in contractual relationships based on collaborative working, in which the design and realization teams share similar values and goals, it may be beneficial to allow some latitude for substitution based on the contractor's knowledge.

Closed proprietary specifications

In specifying proprietary products and not allowing substitution, the designer has made a choice and given the contractor precise instructions on what to use. This is known as a closed specification, since the contractor is not permitted to substitute alternative products for those identified in the written specification. Responsibility for the specification rests with the design organization and not allowing any substitutions is an implied guarantee that the product is fit for purpose and represents good value for money. Warranties, guarantees and insurances should be sought from the manufacturer to transfer the implied liability from the specifier's office. Under traditional forms of contract, the contractor cannot make changes without the permission of the contract administrator. Given that a lot of time and effort will have gone into choosing a particular product in the first place, when a request for a change is made by the contractor many specifiers are reluctant to consider it without a very good reason, for example, a problem with delivery or an unforeseen technical difficulty on site. When changes are unavoidable, care should be taken to acquire and check the manufacturer's warranty before issuing the necessary instruction to the contractor in accordance with the contract. Because many changes are made under time pressures, this is not always done in practice but sought after the event; again, it is not good practice and should be resisted.

Open proprietary specifications

When using open proprietary specifications the proprietary product is identified in the written specification but other manufacturers will be considered if proposed by the contractor. It is necessary to specify the process for evaluation and acceptance of alternative products (e.g. standard clause in NBS). The words 'or equivalent and approved' (or 'or equal and approved') are added to provide some latitude for change and are common in the majority of prescriptive specifications. By adding 'or equivalent and approved', the contractor has some latitude in changing specified products as long as the contractor can demonstrate that the product is equivalent to that identified in the written specification. Approval must be sought by the contractor before the change is implemented.

The design organization remains responsible for the final choice of product and has a responsibility to check that any alternatives suggested by the contractor are fit for purpose and equal to that originally specified. Requests for changes must, under the standard form of contract, allow the contract administrator sufficient time to consider the proposed alternative. There is also a requirement for the contractor to provide the contract administrator with sufficient information (i.e. the relevant technical details and cost) from which a considered decision can be made. Too often, the contractor merely submits a list of products assuming (or hoping) that the design office has the relevant information. If products are unfamiliar to the office, then literature and samples will need to be sought, which takes time and can be arduous. Care should be taken to ensure that any cost savings are fully documented and passed onto the client (not the contractor). The use of 'or equivalent and approved' can lead to arguments as to whether or not the product is 'equivalent' (in practice, some characteristics will be, others will not; hence the arguments). Adding 'or equal and approved' is one way of dealing with the anticompetitive badge given to proprietary specifying, but if a specifier has gone to a lot of trouble to select a particular product, to add such wording is to invite substitution and potential problems. Some specifiers use the phrase 'or similar approved', which is not the same as 'or equivalent and approved' and should be avoided (since this constitutes an open invitation to substitute products which have different properties to those specified).

Some designers specify proprietary products and then add the wording 'or equal'. This is an open invitation to the contractor to use alternatives without asking for approval and is not considered to be good practice. It is used in an attempt to shift liability for product selection from the specifier to the contractor. If the specified product is used, then the specifier's office remains fully responsible, but if the contractor substitutes a product, then liability is transferred to the contractor. By using the words 'or equal' (and not using the word 'approved'), the contractor does not have to seek approval from the contract administrator and can substitute at his or her own risk. And substitute they will. Uncontrolled substitution for cheaper products is a virtual certainty, and cost savings will not be passed onto the client because the substitutions will have been assumed at tender. The danger with using 'or equal' is that clients do not get what they think they are paying for: it is the contractor who profits. An additional concern is the quality and long term durability of the building because substitutions will be made to suit the contractor, who may not be entirely clear as to the original design intent. In an American study of construction claims, litigation or arbitration related to specifications, nearly 25 per cent of the cases were related to the use of 'equal' clauses (Nielson and Nielson, 1981), which was double that of the next most common problem, ambiguous phrasing.

If the term 'or equal' is to be used (and the authors would recommend against it), then the specification should be written to require notice of substitution. This will allow substitutions to be tracked and provide information for an accurate design record of the completed building to be drawn up. There must also be a requirement for the contractor to provide a fitness for purpose warranty. In such situations (of doubt), a performance specification may be a better option because it transfers the actual choice of product to the contractor but at the same time sets defined quality parameters.

Performance specifications

Unlike proprietary specifications, performance specifications do not identify particular products by brand name; instead, a series of performance characteristics is listed for a discrete element of the project (essentially a technical brief), which must be met by the contractor. These characteristics are the attributes required, the desired results or requirements, the measurable or observable criteria, and tests and checks for conformance. It is the design intent (the end result) only which is specified, and so the specifier must take care to describe clearly and accurately the intended result. The contractor must then interpret the design intent and identify products that meet the specified performance and propose these to the specifier. The specifier then has the job of evaluating the proposed solution against the specified performance. Performance specifications provide the contractor with the opportunity to be creative in the way in which the performance requirements are met, the theory being that this should result in a cost effective and efficient construction.

Performance based specifications vary in their scope. They can be used to describe a complete building project, one or more systems, or individual components. For example, clients may produce a performance specification for a design and construct project, engineers may produce a performance specification for the mechanical and electrical specification, and designers may specify such items as fire resistance and thermal insulation by the performance required (which is quantified). A performance specification for a brick would state the required size, colour, texture, durability, water absorption and frost resistance. Depending on the performance standard set, a range of different manufacturers' bricks may satisfy the required performance. This leaves the choice of product to the contractor and is popular in contractor led procurement routes because it gives the contractor flexibility over product selection. Whether or not the contractor chooses Ibstock's Red Rustic (or whatever else the designer may have had in mind) will depend to a certain extent on how tightly the performance requirements have been written.

In passing the choice of product to the contractor, the contractor is given a design function and in doing so is expected to exercise reasonable care and skill in the same way as a designer. The contractor is responsible for the results. A disadvantage to the specifier is that the performance specification may be difficult to enforce. It is not uncommon for the contractor to make last minute changes to products in order to save money or meet programme deadlines. It is also important to check that the product selected complies with the performance specification required; sometimes it does not, the contractor hoping that the specifier does not have time to check thoroughly. Care should be taken to record the final product choice, both as evidence in the event of a claim and for reference purposes for maintenance and when alterations are made to the building.

Performance specifications were pioneered in the USA through their use in a school building programme in California in 1961 and started being used in the UK at much the same time in public housing schemes and the provision of school buildings (Cox, 1994). The Property Services Agency (PSA) was one of the leading advocates of performance specification in the UK, although it recognized that some elements were best specified using the prescriptive method. During this period, it was claimed that performance specifications were an effective way of improving efficiency and encouraging innovation: the reality was quite different, and the use of performance specifications declined rapidly. Performance specifications are best suited to large projects, while on small to medium sized ventures it is common to use descriptive specifications for the majority of the work, with performance specifications for items such as the heating, ventilation and air conditioning (HVAC), where the specifier has less knowledge than the specialist supplier. More recently there has been increased interest in performance specifying, and once again, the argument that it encourages innovation in processes and products and also reduces cost is being promoted, although there is little research to back such an argument.

Performance specifications are generally regarded as being more difficult and time consuming to write than prescriptive specifications. One of the main challenges for the designer is deciding on the level of performance required: too narrow, and the tenderer is given little latitude; too wide, and the scope becomes too great to make sensible comparisons from the solutions presented by competing contractors. Indeed, one of the main disadvantages is that there is no clear and consistent basis for bidding (or evaluating the bids). Care is needed to establish levels of performance that suit the project and the client. Performance specifications tend to be used by client organizations keen to leave the choice open (in the hope of achieving the same performance more cheaply than if a proprietary product were used). One argument put forward is that performance specifications are more effective in ensuring constructability and hence value for money on behalf of the client. There is little evidence to support this claim; indeed, it could be argued that a good design team could ensure constructability and value for money through the use of any of the methods described here.

There are some aspects of building where performance specifying can be simpler than descriptive specifications. The strength of timber depends on both its species and its grade and at one time structural engineers had to specify both to ensure adequate performance. However, in most situations the species was immaterial and in recognition of this there was a shift to specification by strength classes, i.e. the performance of the timber. This left timber merchants free to supply whatever species and grade met the engineer's requirements.

Open and closed performance specifications

The designer's office is responsible for the performance specification because, if met, the solution will be fit for its purpose. The contractor remains responsible for ensuring that the solution meets the performance criteria. Performance specifications are sometimes referred to as 'open' or 'closed', depending on the amount of latitude provided by the required performance criteria. An open (loose) performance specification would be written in such a way as to allow a great deal of freedom of choice for the contractor. A closed (tight) performance specification would be written in such a manner as to limit severely the choice of the contractor, sometimes to one or two possible products. Some specifications are so open as to render them worthless. Where they are so closed (usually because a manufacturer's performance specification has been copied from their technical literature by the specifier) a proprietary approach would have been less time consuming for all concerned.

Price

It is sometimes useful to specify by price, e.g. to specify a prime cost. This is another form of performance specifying, in that the contractor has to meet the requirements within the specified cost.

'Open' specifications

In addition to the four methods described above is the 'open' specification, in which particular items are not specified. Open specifications vary from the unintended (and possibly negligent) to their deliberate and considered use.

Open (silence)

Where there is silence in the documents, i.e. a particular item is not specified, it is referred to as an 'open' specification. This situation usually arises because the specifier has forgotten to specify a particular item. For very minor items, most contracts allow for such situations, and the contractor assumes responsibility by choosing a particular product to suit. An example is that the temporary fixing of timber components is sometimes needed to hold them in place until subsequent operations have been completed; joists fixed to beams until the flooring is nailed down, for example. While locating nails or screws may be asked for, their precise size and length are hardly material and the contractor might use whatever is available.

However, for major items missing in the specification, the specifier's office is probably negligent for not specifying, and responsibility lies with the design office. Clearly, it is difficult to decide what represents a minor item and what represents a major item and caution is required. Careful checking procedures should limit the number of omissions in the documentation. An instruction will be necessary to rectify the silence, and there will be an additional cost to the contract that someone has to pay for.

Open (qualified silence)

Qualified silence is different from silence. An example of qualified silence would be a description such as 'use an approved undercoat' which, although considered, is another way of not specifying, some would argue a lazy way of specifying. Responsibility remains with the design organization as long as the contractor submits details for approval, essentially, a way of delaying a decision and one that may slow down the contractor if approval takes some time to resolve. In the majority of cases, it would be better to specify a proprietary product and allow the contractor some latitude by adding 'or equivalent and approved'.

Client specifications

Some clients, especially those who have a large portfolio of buildings, often develop their own requirements, expressed as a 'client specification'. Client specifications vary in both their complexity and their use of performance and/or prescriptive methods. At its simplest, the client specification will be a list of performance criteria, perhaps supplemented with a list of materials that are not to be used. For example, a client wanting a new warehouse may simply list the floor area and volume required, together with specific requirements for loading bays and percentage of office accommodation to warehousing space. At the opposite end of the scale are client specifications that are extremely detailed. Such lists vary in complexity, depending on the building type and the experience of the client organization. For example, for buildings where hygiene is an overriding consideration, as in food preparation, there will be some specific requirements that must be met. Such specifications represent a source of expert knowledge developed by the client organization over time and revised to suit changing circumstances and improvements to their standard requirements. As such, they provide an excellent briefing document and detailed design guide from which to work. Where design organizations carry out repeat projects for such clients, it is standard practice to develop a bespoke master specification for that particular client.

Lists of products that should, or must, be used and possibly a list of prohibited products also develop from the client's experience, both good and bad. Such lists can be extensive. Although many design organizations take client specifications as a definitive list, some designers question them from time to time, especially in situations where there is discrepancy between what a client wants and what other organizations/control agencies may require to ensure conformity. For example, repair and maintenance work to a listed building may require the use of lead based paint when the client specification clearly states that such paint should not be used. In the majority of cases, differences should be resolved quickly and incorporated into the project specific specification.

The danger of client specifications is that they may not always be aware of the other constraints and this can result in a specification that is impossible for the contractor to meet. In one case the engineer required the use of birch plywood, which would provide the necessary structural performance. The client had also insisted that all timber carried a Forest Stewardship Council (FSC) certificate that guaranteed that it came from sustainably managed forests. The carpentry firm making the components, plywood box sections, found a birch-faced plywood that carried the necessary certification and which appeared to be of good quality. Unfortunately, it came from Latvia and was manufactured for the furniture trade. As a result no one knew its structural properties and the engineer was not prepared to accept it on those terms. The situation was resolved when the engineer pointed out that although the Finnish birch plywood, which was his preferred alternative, did not carry the certification required by the client, the Finns had been manufacturing this product for half a century, which he considered sufficient evidence that their forests were being sustainably managed.

Case study: Using proprietary and performance approaches

As intimated above, there is a number of different approaches to specifying. The approach tends to be determined more by office policy and individual preference than by the professional background of specifier or office. Here, the process is viewed from the perspective of four different specifiers drawn from interviews with those working in different design offices, which helps to illustrate different approaches to specifying. All four approaches appear to work for the specifier interviewed and their design organization.

The first specifier rarely used performance specifications, preferring to specify by proprietary brand name. Any requests from contractors to change specified products were vigorously resisted by this specifier and other specifiers in the design office, simply because it was their office policy to do so. Their argument was that 'the clients should get what they are paying for'. Proprietary specifications were seen as essential in controlling the quality of the building because the materials used were dictated by the designer, not the contractor. This design organization operated a very formalized list of approved products and a list of prohibited products that specifiers had to adhere to. This was part of their quality management system and helped to ensure that (in their opinion) good quality materials were always specified. The feeling in this office was that the designers must choose the products and then stick to them if quality was to be achieved. One of those interviewed summed it up with the comment: 'why bother employing a designer if you are going to let the contractor choose the products, and hence determine the quality?' This particular design office had a reputation for producing good quality buildings that were cost effective. The office used well defined managerial controls, which included the maintenance and use of a master specification and standard details, both of which were regularly reviewed and updated by a senior member of the office.

The second specifier used a mixture of proprietary and performance specifications. She only specified by proprietary name when products were to be seen, i.e. when they formed part of the internal or external finishes of the building, such as facing bricks, roof tiles and internal finishes. Products hidden from view on completion, such as load bearing blockwork, were specified by the performance method. Her particular design office did not have a standard policy on specification that all designers followed; instead, there was a wide variation in individual preference for specifying. Other designers within this organization also varied their use of performance and proprietary specifications to suit the project and their client. There was no office dictat on specification writing; indeed, the whole office was very loosely managed. There was no office master specification; the universally adopted habit in the office was to roll project specifications from one project to the next. This office also had a good reputation for the quality of its buildings, but was also well known for its inability to finish a project on time and frequently over spent the budget. The specifier claimed to know that rolling specifications was not good practice, but that it 'worked' for this particular office.

The third specifier, as a matter of office policy, specified entirely through the use of performance specifications. He rarely specified by brand name unless forced to do so, leaving the choice of product to the contractor and subcontractors. Proprietary specifications were sometimes used to suit project specific requests from town planning officers and in response to client requirements. Because the performance specification was written for the majority of the elements specified by this designer, it was easy to draw from the master specification. Not only was this quicker for this individual to implement, but he was keen to keep building costs down for his clients. He believed that this method allowed the contractor to select the cheapest products that met the specification, and proprietary products would have been more expensive. This office was engaged in a significant number of contractor led projects (design and build, management contracting) and the contractors they worked for preferred the performance method to proprietary specifications, thus allowing some latitude in product selection. It was acknowledged by the specifier that working closely with contractors influenced the way in which they specified.

The fourth specifier worked in a large organization that had set procedures for specifying, although they were not followed. According to the office manual the specifiers had to specify by proprietary name (unless otherwise agreed with the client) and product substitution should be resisted. However, the specifier claimed that it was standard procedure for specifiers to encourage the contractor to change products at the tender stage in an effort to save money. This made it look, on paper at least, as if they were saving the client money because of the difference between the initial cost estimate and the tender sum. The clause 'or equal' was used in the written specification, and the contractor was encouraged by the design office to take full advantage in order to reduce costs for the client and maximize profits. The priority for this designer (and his colleagues) was to keep the initial cost of the building as low as possible; his attitude was that one product was much the same as another, and thus substitution by the contractor was not an issue. He was aware of the fact that some of the products substituted may need to be replaced sooner than those initially specified, but since his client sold the buildings on completion, it was not his, or his client's, main concern. There was little documentary evidence to show what products the contractor had or had not changed, nor was there any real effort to pass the cost savings to the client. Rolling specifications from one project to the next was common practice. At the time of the interview, the office was having difficulties with the quality of their finished buildings and was faced with a number of litigations.

In addition to illustrating different approaches to specification, these four examples demonstrate differences in approach by different offices to the management of design. Some specifiers hold very strong views on how to specify, as do some design organizations, whereas others are more relaxed in their habits and procedures. When conducting the interviews, there was a certain amount of surprise that someone (even a researcher) should be interested in the specification process. As one interviewee observed:

> Specifying is something you have to do very quickly, there is little time to think too long about the whys and wherefores; the only time we really give it much thought is when there is a problem.

None of the specifiers interviewed felt that the specification process was pivotal to good quality buildings, maintaining that other tasks (such as detailing) were more important (a view consistent across samples of architects, technologists and surveyors). The authors beg to differ, especially given the amount of poor practice found in the course of the interviews. There is a suggested relationship between specification practice and the resulting quality of the various offices' products. However, more research is needed to substantiate such a link.

End of chapter exercises

- Take a small part of your current design project and identify three different products or components. Which specification method would be best suited to each of your selections, and why?
- Discuss the approach taken by each of the specifiers described in the case study above. What are the advantages and disadvantages of each approach as viewed by: (1) a specifier; (2) a design manager; (3) the contractor; (4) the client?
- Select a product type where you think that the products available on the market are unsatisfactory. Assume that you are going to approach a manufacturer to develop an improved version. Draw up a detailed specification from which to begin this process.

3 Information sources

Professionals have a duty to stay up to date with current regulations and codes, current building practices, changes to forms of contract, and developments in materials and products, both new products and those rendered obsolete. Although this may sound a relatively easy thing to do, in practice it presents a series of challenges. The volume of information available to the building designer is vast, each new project bringing with it a new set of challenges and a fresh search for information to answer specific design problems. One of the problems facing many designers is that they do not need to access the information sources all the time. Indeed, many designers work on many different stages of jobs and different projects concurrently, and the physical act of selecting products and specification writing does not take up a great deal of their time. This is particularly so of designers who run a project from inception to completion, mainly those self employed and working in small offices and/or on small projects. For these individuals, access to information to assist them with the specification writing may only be required every twelve months or so, and then only for a few weeks until the task is complete: they therefore need a reliable and current source of information that can be accessed quickly.

Organizations that subscribe to one of the information providers, such as the Barbour Index or National Building Specification (NBS), are kept up to date with the majority of new developments that affect specification writing. If not, they will have to rely on reading about changes in the professional journals and trade information. In practice, the majority of practitioners try to stay up to date through information from a variety of sources. Knowing where to find a particular piece of information in a crisis is one of the prerequisites of staying in a job. More specifically, the specifier needs to keep up to date with:

- *Building Regulations and Codes*: subscription to one of the on line information providers will ensure that the regulations and codes accessed during the design and specification process are current.
- *Building practices*: staying up to date with current building practices is a little more problematic. In part, this is because it is rare for two designers to agree on the best way of detailing and specifying a building.

- *New materials and products*: manufacturers are constantly seeking to improve their products and expand their market share; thus, products may be 'improved' or replaced as part of their strategy. Specifiers have to keep up to date with these developments in order to specify effectively. Another problem is keeping up to date with materials and products that may no longer be available, perhaps because of safety concerns or simply because the product was not commercially viable.

For designers dealing with existing buildings, the challenge is reversed. Their problem is to find the relevant codes and building practices that existed at the time the building was constructed (and or remodelled). To ensure satisfactory performance of the structure and building envelope, it may also be necessary to match the properties of the new work with that of the existing. Yeomans (1997) has clearly demonstrated the difficulties for the practitioner in finding information on early building products. As noted in the last chapter, it may be necessary either to carry out a search of early literature or to undertake tests to determine the properties of the existing materials before the specification for new work can be written.

The individual designer cannot, and should not be expected to, survey the whole body of literature available. Instead, an easily accessible, accurate, concise yet comprehensive body of information is required. The main sources for designers working in Britain are:

- Building Regulations and Codes
- British (BS), European (EN) and International (ISO) standards
- British Board of Agrément (BBA) certificates
- Building Research Establishment (BRE) publications
- manufacturers' technical literature
- compendia of technical literature, e.g. *Specification, RIBA Product Selector, Barbour Index*
- trade association publications
- technical articles and guides in professional journals
- previous projects worked on by the organization
- office standard details and master specifications
- building information centres

Research by Mackinder and Marvin (1982) found that building designers tended to refer to written documentation only when they had to, preferring to rely on rules of thumb and their experience until such time that they were forced to search for information. Clearly, there comes a point when someone with technical knowledge will have to make the design work, including compliance with legislation and best practice. In some design offices, this is carried out by the same person who crafted the conceptual design; in others, the task of detailing and ensuring the design complies with prevailing legislation will be carried out by more technically orientated individuals, technologists and specification writers. Although this separation occurred during the period when information was paper based, there is nothing to suggest that it is any different since the electronic revolution. Indeed, one would expect designers' behaviour to remain much the same.

In the past this information was held in the design office library as paper copies, but more recently, many design offices have moved towards digital information sources. Information was initially held on CD-ROM, but now more commonly it is accessed on line, either directly via the manufacturer's homepage and/or via subscription to an on line information broker, such as the Barbour Index. This makes access quicker, and the material should be up to date and reliable in content. There is a large number of interrelated sources of information that the specifier can draw on, explored below.

Regulations

In addition to satisfying the requirements of the client, the designer needs to adhere to standards and regulations established to ensure public safety, or some other communal good such as energy conservation. The designers will commonly refer to these standards during the design process so that, although they are not produced primarily as a source of design information, they serve much the same function. The extent to which they do so depends on the degree to which the standards and regulations are prescriptive, a factor which has changed over the years (see below). The same is more so of design codes, most commonly used in structural design, where in addition to setting such parameters as the required design loads, these codes commonly set out the method of design providing formulae for the designer to use. In such cases the codes of practice are a vital information source, being used more or less as handbooks by the designer. But they will also establish standards of workmanship to be applied, since the performance characteristics of the final construction will depend on these.

Prescriptive and performance specifications both rely heavily on reference to current regulations, codes and standards. Building Regulations and Building Codes are first and foremost concerned with ensuring safe buildings and providing a healthy environment for building users. Regulations, in whatever form, offer designers a familiar set of controls to work with. They also offer a series of constraints based on experience, research and common sense – essentially a guide to best practice. National regulations, such as the Building Code of Australia or the Building Regulations, *must* be complied with. Standards, such as International Standards that have been adopted by national authorities, and codes, such as design Codes of Practice (CPs), *should* be complied with. Busy designers may well view regulations, codes and standards as a burden because they are time consuming to read and act upon. There is always the danger that a standard is referred to in a specification without the specifier necessarily reading the standard and/or fully understanding its subtle complexities. Nevertheless, one of the many skills of the building designer is his or her ability to keep up to date with current legislation, applying it in a creative and cost effective manner to realize the design intent.

Regulatory frameworks

Regulatory frameworks vary between countries, and regional variations are not uncommon in many countries. For example, in Australia, there are state and regional variations to the Building Code of Australia to accommodate specific regional conditions. Globally, the trend has been to move towards a performance approach in preference to a prescriptive one. Prescriptive regulations show or describe the construction required to achieve conformity. Performance based regulations stipulate a level that must be met (or bettered), but do not specifically indicate how this is to be done. Another difference between prescriptive and performance regulations is that performance based regulations focus on methods of conformity that consider buildings as whole systems, rather than elements in isolation. Thus, trade offs between component parts, for example in achieving the necessary minimum thermal insulation values, are allowed to achieve the given objectives. Theoretically, the performance approach allows designers greater freedom of expression in the way in which conformity is achieved, although in practice there is still a heavy reliance on standard details (usually copied from the regulations) to save time.

In the UK building was controlled by local legislation, the building by laws, which were established under the Public Health Acts of 1875 and 1890, until the first national Building Regulations were introduced in 1966, which brought all of England and Wales under a common legislation for the first time, replacing the varying local by laws. However, some types of building were also controlled under other legislation, such as the Shops and Offices Act. The 1966 Building Regulations took their form from the prescriptive nature of the earlier building by laws, which sometimes had the effect of hindering innovation in construction. In England and Wales the Building Act of 1984 led to redesigned regulations and a set of *Approved Documents* in 1985 that described construction that met the requirements of the Building Act, but otherwise left designers free to produce alternatives. This was a major shift from a prescriptive approach to a performance-based one. The *Approved Documents* are intended to provide guidance for some of the common forms of construction while encouraging alternative ways of demonstrating compliance under the 'deemed to satisfy' standards.

In Northern Ireland construction is covered by the Building Regulations (Northern Ireland) 2000 with *Approved Documents*, similar to those for England and Wales. In Scotland the Building (Scotland) Act 2003 has replaced the old prescriptive standards with performance standards. This Act is a response to European harmonization of standards and their use in Scotland as required under the Construction Products Directive (CPD). This act also aims to provide designers and engineers with greater flexibility in meeting minimum standards.

Designers and builders now have a choice: they can accept the suggested method in full, in part, or not at all if they can demonstrate an alternative method of compliance. In reality, many designers and builders find it quicker, easier and more convenient to work to the solutions suggested and illustrated in the *Approved Documents*; alternatives are more time consuming and may well be rejected, leading to delays. Since their redesign in 1985, there have been several revisions and additions to the *Approved Documents*, with a major revision in 1991, and ongoing updates. In their current form, the *Approved Documents* form a useful *aide mémoire* and detailed design guide.

Standards and Codes of Practice

Standards and Codes provide guidance to designers based on best practice and research findings. They are an effective way of bringing research and development to practitioners, thus aiding the adoption of new technology and raising standards of quality. Standards and building codes are prepared by committees of specialists drawn from government, academia, manufacturing, professional practice and user groups. The results are documents arrived at by consensus and, as such, they may not meet (or indeed be seen as relevant to) the needs of all. National research organizations such as the BRE in the UK and international research bodies such as the International Council for Research and Innovation in Building and Construction (CIB) are actively involved in the development of national and international standards through representation on various development committees. For example, the CIB has been active in the development of standards for sustainable construction through the International Organization for Standardization (ISO).

Standards and Codes of Practice have two functions. On one level, they provide the designer with advice and guidance, and on another, they provide the specifier with a certain amount of security since they represent best practice. Working to both relevant and current standards, the designer will be safe in the knowledge that he or she is applying the most current knowledge. This reduces risk for the design firm, ensures the safety of those doing the construction and protects the interests of the client. However, it should be remembered that many standards are developed in the light of failure, and problems may, unfortunately, still occur. Structural codes are reassessed and sometimes revised in the event of a structural collapse. Designers working at the cutting edge of technology are likely to be ahead of the relevant standards since they take a long time to develop and/or revise. Dangers can also arise when designers are, without realizing it, working outside the limits implicit in the drafting of the Code or Standard.

It is instructive to consider some of the ways in which the development of standards might be linked to innovations in building. The shift from hot rolled steel sections, used for roof purlins in factory building, to cold rolled sections led to a number of collapses and the realization that the current Code of Practice for snow loading was inadequate. This had failed to allow for drifting of snow in multiple pitched roofs, but the uncertainties inherent in the design of hot rolled sections had resulted in sufficient built in margins of safety for this not to have caused problems. Cold rolled sections could not be so easily designed and manufacturers provided 'safe load tables' based on testing. The result was a more accurate estimate of the load that could be carried and hence a lower margin of safety that failed to cope with actual loads that were greater than the design loads. The result is that we now have a more sophisticated method for assessing snow loads.

New standards and codes

The development of new but similar products by different manufacturers results in a wide range of properties that can be confusing to the specifier. Nationally or even internationally applied Standards address this problem. The British Standards Institute (BSI) was the first national standards body, but now there are more than 100 similar organizations, which belong to the ISO and the International Electrotechnical Committee (IEC). Formed in 1901 (as the Engineering Standards Committee), its first British Standard concerned with building was published in 1903, standardizing the sizes of rolled steel sections (BS4). Licences are now given for products to carry a 'Kitemark', which certifies that the product complies with the relevant British Standard. The first of these was issued in 1926, and British Standards have become an essential tool for building designers.

In 1942, the British Standards for Codes of Practice (CPs) for design were introduced to ensure a degree of uniformity. By designing in accordance with the CPs, designers were 'deemed to satisfy' the legal requirements of the time, and many became standard works of reference (Yeomans, 1997). Codes of Practice are based on a combination of practical experience and scientific investigation, and form an essential part of quality assurance schemes. Since the early

1990s, Eurocodes have been published that establish standards for the design of structures across the European Union. Examples include Eurocode 5, *Design of timber structures* (1994) and Eurocode 8, *Design provisions for earthquake resistance of structures* (1996).

With increased attention on the international market has come a focus on International Standards. The ISO was founded in 1947 with the aim of harmonizing standards internationally. As with national standards, such as ASs (Australian), BSs (British) and DINs (German), the ISO series serve as guidance to designers and specifiers and do not necessarily have to be complied with. National standards are being replaced with European Standards (ENs) and ISOs, where appropriate; for example, the ISO series on quality, BS EN ISO 9000–9004, has superseded BS 5750. Likewise, the Australian and New Zealand codes are being replaced with ISOs. This has an advantage in those sectors of building that are attracting firms of architects and engineers who operate across national boundaries.

Although standards and codes are in a constant state of development, the conservative behaviour of designers may result in their failing to keep up with developments in the latter. Keeping abreast of changes in standards is a relatively simple matter compared with keeping up with design codes. Learning a design method represents a considerable investment on the part of the designer. The result is often considerable resistance to the introduction of a new code of practice because this will mean that designers will have to learn new routines. Figures that the designers carry in their heads will no longer be relevant, and even the whole philosophy of design may have been changed. This is what happened when structural codes changed from a reliance of allowable stresses to load factor methods. Thus, where old codes have not been withdrawn, some designers persist in their use. Some structural engineers are still using BS 449, *The use of structural steel in building*, even though a new code has been available for many years.

Trade associations

The trade associations can usefully be divided into organizations representing general materials, such as the Timber Research and Development Association (TRADA) or the British Cement Association (BCA), those representing a type of product, such as the Brick Development Association (BDA), and those representing groups of specialist contractors, such as the National Federation of Roofing Contractors. The first two types will produce information that is useful at the design stage. For example, TRADA produces Wood Information sheets that provide information on general issues such as the design of laminated timber or on the design of timber based construction such as fire resisting walls or separating floors. Some even publish standard specifications. The intention here is to help the designer at this important stage in the process, improving the prospects that the material whose interests the association represents will be incorporated into the design. The designer's concern at that stage will be the performance of the component or assembly, and so it should be this information that will be found in the trade association literature.

A quite different function of trade associations is to provide an assurance of construction quality. This may be done through a wide range of quality assurance schemes. The simplest is perhaps the regular testing of manufacturers' products or the monitoring of their quality control procedures, the advantage to the specifier being that the trade association offers a measure of independence and hence reliability. Examples of this are the TRADA schemes for trussed rafter roofs and fire doors, where products assured within the schemes carry identifying marks for simple checking on site. Associations of specialist contractors may limit their membership to those who comply with certain standards set by the association, for example standards of training for their operatives. The assurance being offered here is that the work will be carried out in a satisfactory manner, something that may be difficult to check on site once the work has been completed or that may even require specialist knowledge to assess. Of course, the judgement that must be made by the specifier is whether or not the claims made by such associations are delivered in practice. Some years ago, the literature of one association implied that it employed a particular academic as a consultant. On enquiries, the person named denied any such link. In spite of this, such associations do perform an important quality control function. In the UK, the majority of house builders belong to the National House Building Council (NHBC), which sets its own building standards (some of which are more stringent than the Building Regulations) and training standards for members. In recent years, the organization has made considerable progress in improving the quality of housing built by their members, for which new owners receive a guarantee.

The larger trade associations are usually represented on British Standards committees and will have an input to changes to the Building Regulations. They also have an input to the development of the Eurocodes.

Property is a major asset, and it should come as little surprise that the larger insurance companies exercise an influence over building standards. In addition to the NHBC's insurance scheme, many large building projects are vetted by insurance companies at the design stage to assess the amount of risk against their own guidelines for security and fire protection. Other organizations work to design guides such as the Housing Association Property Mutual *HAPM Component Life Manual* (HAPM, 1991), which gives extensive information and benchmarking for component service lives of materials and some mechanical and electrical (M&E) components. In France, the requirement for buildings to be insured means that insurance companies require that designs are checked, effectively taking on the role performed by building control in the UK.

Testing and research reports

Independent research and test reports published by recognized building research organizations and papers contained in peer reviewed academic journals are the best source of information, but are seldom used by practising designers as they are not written with this audience in mind.

For some products, such as thermal insulating materials, the Building Regulations require all new products to be certified by an independent organization before they will be approved by a local authority. An Agrément certificate, British Standards Kitemark and or the Conformité Européenne (CE) Mark are therefore essential for companies hoping to sell their products.

The British Board of Agrément

An Agrément Board was set up in Britain in 1966, modelled on the French Government's agrément system that already had an established track record. In 1982, the Agrément Board became the British Board of Agrément (BBA). This is an independent organization that is principally concerned with the assessment and certification of building materials, products, systems and techniques. The status of agrément certificates is defined in the *Manual to the Building Regulations* 1995. Certificates guarantee compliance with the regulations where health and safety, conservation of energy and access for disabled people are concerned. Specifying a product or system that carries an agrément certificate will give assurance that the system or product, if used in accordance with the terms of the certificate, will meet the relevant requirements of the Building Regulations. Products assessed by the BBA are usually new to the market or are established products being used in a new, or innovative, way.

British Standards Kitemark

Kitemark schemes were introduced in 1902 and now cover a variety of products and services. This is one of the most respected product and service certification marks available, and many organizations require a manufacturer's product to comply with the Kitemark before it can be specified and/or purchased. The BSI provides a wide range of testing services for construction products, from product development to independent testing of a product against a technical specification or standard (e.g. BS, EN, ISO and Trade Association specifications) and testing for Kitemark certification.

CE mark

Manufacturers wishing to place their products on the market in any of the member states of the European Economic Area (EEA) must apply CE marking to their products. A range of compliance routes is listed under the *New Approach Directives*, which in most cases will require a manufacturer to use a notified body to assist with certification. Like the Kitemark scheme, CE marking provides the specifier with some reassurance that the product has been independently tested and meets European standards.

Building Research Establishment

The BRE is one of the most highly respected research organizations. Originally set up as the Building Research Station and government funded, this organization now operates on a commercial basis. The BRE publishes a range of informative material on a wide range of technical issues relating to construction. These include the BRE Digests, BRE Defects Action Sheets, BRE Good Building Guides, BRE Information Papers and BRE Reports. It also holds a significant amount of information relating to all issues of building in its library. It has been, and continues to be, a comprehensive and reliable source of information for designers.

The BRE has also been instrumental in developing a number of tools and guidance documents on environmental sustainability. The BRE Environmental Assessment Method (BREEAM) helps the design team to measure the environmental sustainability of both new and existing buildings. This evaluation tool takes into account a number of factors, such as energy use and indoor climate, and includes building materials. The BRE has also published *The Green Guide to Specification* (Anderson et al., 2002), which provides information on the environmental impact of categories of materials. These are rated by 'Ecopoints', numerical values derived from life cycle assessment studies of environmental impacts of a range of materials and components. This allows the specifier to make direct comparisons, for example between steel and concrete, for the same element of construction, which would not be possible from the manufacturers' literature alone.

Manufacturers' own standards

Manufacturers set their own standards for production, delivery and after sales service. The majority of manufacturers work to the current standards, simply because if their products do not comply with these, they would not be specified. Some manufacturers set higher standards than those within national and international standards, because they have the manufacturing expertise to do so. Combined with installation by their approved fitters, they can guarantee standards of performance and provide the necessary warranties and insurances that specifiers require. The results of this work may not be contained in research reports that are readily available, but will sometimes form the basis of tabulated performance figures contained in their literature, such as 'safe load tables' for structural members.

Manufacturers' information

Collectively, building product manufacturers produce a variety of information for a variety of uses, from promotional literature to technical literature. In many respects, the pattern books of the late nineteenth century and the manuals on building construction of the twentieth century have been replaced by manufacturers' own technical literature that often contains full working details showing the way in which their products can be used. This is a necessary feature of the shift from largely traditional forms of construction to the use of modern, and often highly innovative, building products. Manufacturers' information can even include typical specification clauses that can be easily used by designers. At the other end of the scale is promotional literature that is simply designed to raise awareness of the company and its products to the specifier. Rarely does it provide enough information to allow the specifier to specify the product; rather, the intention is that the specifier should make contact with the manufacturer to ask for further information. The manufacturer may then choose to send technical literature, deal with enquiries by telephone and/or send a technical representative to the specifier's office to assist with the specification. While the specific function of different kinds of manufacturers' information may be different, all raise the awareness of the specifier to the products described and, therefore, the general term 'trade literature' is used here for all.

Trade literature

Trade literature is designed with the principal objective of helping manufacturers to increase their sales, and the consistency and quality are as varied as the products on offer. Building materials and products manufacturers are in business to sell their products to specifiers, and to do so, they must make them aware of their product range. From the earliest days of mass production, manufacturers have advertised their products to potential specifiers, the designers, builders and tradespeople, through trade literature and advertisements in the specialist journals. Product catalogues were, and still are, an effective means of selling products because they are convenient for specifiers to use and in effect order from, through the act of specifying the products described in the contract documentation. The recent provision of typical details and specifications on disk, or more recently from the Internet, which can be imported into the construction details and specifications, has been developed in tandem with the general adoption of computer aided design (CAD) in architects' offices. In many respects, information available through the Internet is not significantly different from that offered in printed form; the difference is the ease with which it can be transferred to contract documents. There is evidence that paper copies (received in the post) are also retained by the majority of specifiers, and in some cases preferred to digital sources (Barbour Index, 2006). There appears to be a pattern of behaviour, in that specifiers access on line information as part of their initial search for information and then request paper copies of information from manufacturers when they start to consider the properties of the product in more detail.

By making it easy and quick for the designer to import their standard details and standard specification clauses into project information, they can find their products specified without even being contacted by the designer. This 'free' information has been carefully designed so that by selecting a particular manufacturer's detail, the designer is also confirming his or her choice of product, so that the provision of a proprietary product specification is an effective method in helping the designer to adopt a particular product over that of a rival manufacturer. This is true of proprietary specifications and performance specifications, the latter being written in such a fashion as to make the selection of a rival manufacturer's product impossible. For example, a performance specification for a particular product may specify a manufacturing tolerance that rivals cannot achieve. In the longer term, manufacturers hope that the designer, and hence the design office, will adopt their particular detail and their product as a standard detail. Many manufacturers will also employ technical staff who will provide bespoke details for a particular project.

A feature of manufacturers' own details is that their literature may well contain errors. Manufacturers are concerned with the promotion of their own product range so that other components may not necessarily be represented correctly. A striking example was the way in which trussed rafter roofs were introduced to Britain. Manufacturers' literature tended to emphasize that these were engineered products, which was true. However, unlike American roofs for which they had been developed, British roofs might need to include chimney openings and, at that time, cold water storage tanks. The promoters' brochures suggested ways in which trusses could be trimmed to accommodate the former while, according to the trade information, bearers could be placed on the tie beams to support the latter. The aim was presumably to suggest the ease with which the new product could be incorporated into existing building practices, the writers of these brochures not apparently noticing that their recommendations contradicted the idea of an 'engineered product' (Yeomans, 1988b). As with all 'typical' details, caution should be exercised when working such details into the overall detail design drawings, because, once included, they tend to remain.

There are occasions when the manufacturer may not have the information required. In one surprising example the architect's design called for a folding door and, as the supporting structure was a rather long span timber beam, the engineer was concerned about deflection limits. With the door half folded there would be a considerable load at mid span and it was important that the bottom guide did not foul or the door would stick. The manufacturer's technical information said nothing about the maximum deflection of the supports, nor was their technical department able to help. Eventually the engineer had to call for detailed drawings of the door guide and determine the deflection limit for himself.

Filtering manufacturers' information

Research published in 1971, *Architects and Information*, by Goodey and Matthew, looked at how architectural firms handled technical information, with a view to designing information that might appeal to specifiers. At the outset of their research, it was known that much effort was being wasted because information produced (both technical guides and manufacturers' trade literature) was not being read by architects in practice. They concluded that this was because it was not provided in an accessible form for specifiers and suggested a number of design guidelines to be followed by manufacturers. In the event these recommendations were largely ignored by manufacturers, despite the existence of British Standard 4940, *Technical information on construction products and services* (BS, 1994).

Although there has been a shift to digital transmission of technical information, the fundamental need for accurate, reliable and informative information has not changed. Unfortunately, it does vary in quality and, as such, needs to be controlled by the specifier's office to avoid information overload and to ensure that the sources that are used are reliable. The way in which it is controlled is an aspect of office management policy, usually applied at a senior level. This has the effect that the more junior members of staff in medium sized and large practices are shielded from excessive information or denied knowledge of potentially useful innovations. Whichever way one looks at it, this filtering practice must affect the specification process and the adoption of building product innovations.

Information overload occurs when an individual or organization receives more information than it can handle, leading to some form of breakdown (Rogers, 1986). It is tempting to view information overload as a new phenomenon, but it has been a cause for concern amongst professionals for some time. For example, in *The Principles of Architectural Design*, published in 1907, Marks suggested that architects should look at the trade catalogues left by travellers (the forerunner of the trade representative) on a weekly basis and then dispose of them after reading – an early example on managing the volume (if not quality) of technical information. The volume of information targeted at specifiers has grown significantly since Marks' advice. Goodey and Matthew reported that design offices were being 'submerged by a flood of literature' in their report published in 1971, a situation that has worsened with the advent of digital technologies. Studies in Japan (e.g. Bowes, 1981) and the USA (e.g. Pool, 1983) into information use by societies in general found an extraordinary growth in information available, yet a modest growth in the consumption of information, suggesting that individuals operate selective exposure, which is discussed from the perspective of the specifier in Chapter 8.

Filtering of information is known as 'gatekeeping' in communication studies, a metaphor first coined by Lewin (1947), studied empirically by White (1950) and since developed into a field of research most commonly concerned with mass media studies. The gatekeeping metaphor can be applied to any decision point where information, provided via mass media or interpersonal channels, is assessed for onward transmission by a 'gatekeeper' (Shoemaker, 1991). The construct is present in diffusion of innovation literature, both explicitly (e.g. Greenberg, 1964) and implicitly (e.g. Rogers, 1962) and is present in literature concerned with knowledge acquisition, where the term 'technological gatekeeper' is used to describe an individual who attempts to control information entering the organization from external sources (e.g. Leonard-Barton, 1995). Such individuals operate at the boundary of the firm, where they 'browse' information for its relevance to both themselves and their organization's members, withholding, altering and transmitting information as it passes into the social system over which they have a certain amount of control. Thus, gatekeepers may influence the innovativeness of the organization, simply by the type of information they allow through the gate.

Professionals are personally responsible for the advice that they give and in offices other than sole practitioners this responsibility falls upon the partners. Therefore, they will manage the office in a way that minimizes their exposure to risk and this includes exercising some control over the products that are specified. The first stage in this process is often to control the product information coming into the office, i.e. operating a gatekeeping mechanism. The partners and the office receptionist are potential gatekeepers, being the first gate encountered by trade representatives and trade literature. The trade representative must pass the office receptionist to see a potential specifier, and direct mail must survive any filtering by a partner when the mail is opened if it is to stand any chance of finding its way to the specifier.

Emmitt (2001) investigated how and why trade literature was managed as it entered architectural offices. The partners turned out to be ruthless in their handling of trade literature, as the majority (80–99 per cent) was thrown away during the morning ritual of opening the mail. On average, the partners said that approximately five minutes was all the time that they spent on this task, and most said that the literature had only three to five seconds to command their attention, otherwise it was thrown away without even looking past the front cover. Two factors emerged that appeared to lead to retention: first, whether the literature was relevant to the type of work of the office, and second, whether it appeared to contain enough technical information to make it worth keeping. Literature produced by companies that were familiar to the office was more likely to be let through the gate than that from companies less well known to the partner. This was seen as a risk management technique and was consistent across the full sample, regardless of office size. Interviews revealed an underlying level of dissatisfaction with the general standard of trade literature.

In the very small offices, the partners also acted as specifiers because they claimed to retain a 'hands on' approach, dealing with design issues in addition to running the business side of the practice. Here was a difference between the very small offices and the others because the partners in larger offices were primarily concerned with management issues; none of them was personally involved in product selection, despite the fact that they claimed to be closely involved with all of the jobs. Of course, one cannot be sure of this, because Cuff (1991: 20) noted the differences in partners' actual behaviour and their own self image. However, their reported behaviour was reflected in office procedure, where trade literature was passed from the senior partner to a partner or associate who was responsible for technical issues. Thus, once past the first gatekeeper, the information faced further filtering by a less senior member of the office: the estimated reduction in literature at this stage was a further 50 per cent, based on an objective (and more time consuming) assessment.

One of the medium sized to large practices studied by Emmitt (2001) employed a very elaborate gatekeeping system. Any trade information let through the gate by the partner went straight to a senior architect (associate level) whose job was to deal with technical matters in the office. If the product was perceived to have a degree of benefit to the practice, he then contacted the manufacturers directly for further information. Once this information was received (usually delivered by a trade representative), the literature was either discarded, if unsuitable, or presented to a partner's meeting and, if deemed suitable, recommended for adoption onto the practice's list of approved products. The reason for adopting such a complex system was founded on concern about product failure and the threat of legal action that might ensue, i.e. it was a risk management system. All staff had to select from the office palette of approved products and were indirectly being restricted in their awareness of new products by the gatekeeping process. There is no indication of how common this particularly elaborate process might be; it would only be feasible in a reasonably large office. Nevertheless, the general principle of restricting the kind of products that employees may use in order to limit risk is very common, although not all will apply the restriction at this 'awareness' stage.

To a large extent, partners of architectural offices assumed that their experience and professional judgement would help to prevent the specification of unsuitable products by other members of their office, by limiting the specifiers' choice of products from which to specify: risk management was the primary objective. The specifiers who actually selected building products were selecting from filtered information and so were affected by decisions taken by the gatekeepers because their access to, and awareness of, information was restricted.

The common practice is then for this filtered information to be placed in the office library. Although many offices are working with digital information systems, many still retain a small office library of paper based manufacturers' literature. The Barbour Index (2006) reported that this relatively small collection of information had been selected by the office members for its relevance. Thus, it would appear that the gatekeeping tendencies observed by Emmitt (2001) are still being operated.

Journal advertisements

Manufacturers have also tempted designers to use their new products through advertising in the professional journals. The first and subsequent editions of journals such as *The Builder* (1843) and *The Architects' Journal* (1895) have carried advertisements from a wide range of manufacturers selling an ever wider range of products and services. Advertisements are important for the journals, since the revenue generated by them helps to finance their production and distribution. Some of the product journals in the UK, such as *What's New in Building* and *Building Products* that exclusively contain advertisements are distributed free of charge to specifiers' offices because the entire cost is borne by advertisers. Specifiers need to be aware of this and to exercise a degree of caution in using technical features in journals because many are little more than a rework of press releases of manufacturers, usually evident from the lack of any critical discussion of the components being described. Both forms of advertising are used with the intention of raising specifiers' awareness of the company's product. If this happens, the specifier becomes aware during a 'passive phase', i.e. not actively looking for information about building products.

Once the specifier has become aware of a building product through a journal advertisement, he or she can telephone the company to ask for their literature, fill in a 'reader reply card' (if provided by the journal), or e-mail the company to request further information. Any of these ways will trigger a mailing of literature to the architect's office (by post or via e-mail) that is often followed by a telephone call from the company's marketing department or a visit from a trade representative (if the manufacturer employs them).

Product compendia

Typical examples in Britain are the *Architects Standard Catalogue (ASC)*, the *Barbour Compendium*, the *RIBA Product Selector* and *Specification*, while in America, *Sweet's* catalogue serves this vital role in architects' offices. *Specification*, which was first published at the end of the nineteenth century, sets out to provide guidance on specification. Set out under different trades, it describes the processes of construction and provides specification clauses that specifiers could use.

Compendia are a compilation of individual manufacturers' building products that are published annually. They provide a convenient and familiar point of reference for busy specifiers. These list products under general subject headings and to differing degrees include advertisements by manufacturers, but do not cover all manufacturers, only those who pay to be included. The compendia, available in both paper and electronic format, are not designed to offer advice on product selection, nor do they provide a comparative assessment of similar products; they merely list building products and provide generic descriptions of materials. There is no comparative advice on cost or performance. Thus, the specifier cannot refer to a publication that provides comparative product assessment (unlike, for example, the potential car purchaser, who can refer to specialist journals that provide comparative information about cost, performance and value for money). They do, however, provide a useful source of information, and the interactive web based compendia allow the specifier rapid access to manufacturers' own websites and additional technical information.

Compendia are intended as a source of reference for the specifier, so awareness through this medium could occur a long time after the first advertising campaign has finished. This form of awareness relies on the potential specifier looking through the compendium and then contacting the manufacturer for further details, i.e. the specifier has to be searching for information. While the compendium will not draw attention to the newness of the product, in comparison to the advertisement, it may be perceived as new to the specifier who is looking through the directory.

Manufacturers' websites

Information is also available in electronic format, on computer disks, on CD ROM and, since 1998, freely available on the world wide web via manufacturers' homepages. Manufacturers hope that by providing typical details that include their product in a digital format, which can be easily imported into the architect's drawings, there will be a greater chance of specification.

All these forms offer benefits in speed of response from a specific manufacturer for additional information and save space taken up by paper trade literature in the office library. Although the Barbour Index has identified a clear move towards the use of digital information, it is still common for specifiers to use both paper and digital sources (Barbour Index, 2003, 2006).

According to the 2006 Barbour Report the main reasons for visiting a manufacturer's website are to establish a product's suitability for a project, to download literature, to find a telephone number and to search for a specific product from the manufacturer's range. This tends to suggest that the specifier is already aware of the manufacturer and product range. After visiting the website the usual action is to print out the information from the website and then contact the manufacturer by telephone for further information.

Some comment is appropriate here on information available through the web. There has not been the opportunity for systematic research on the nature of and response to this form of information, so we can only rely on anecdotal evidence at this stage. Nevertheless, it is clear that standards of websites vary as much as printed literature, with the added difficulty that some sites are not easy to navigate. A difficult to navigate site that eventually produces no useful technical data is unlikely to encourage the use of that product. Similarly, a site that requires one to register and log in before being able to access the information is likely to discourage the specifier and encourage further searching for a rival product. These concerns were voiced by specifiers in the 2006 Barbour Report. However, specifiers were not put off investigating the manufacturer further and resorted to telephoning the manufacturer for additional information, hence creating an interpersonal communication channel.

Manufacturers' trade representatives

It would be misleading to give the impression that designers develop their detail designs and specifications in isolation. Successful design relies on cooperation between manufacturer and specifier. Manufacturers have a vital role to play in helping the designer to detail particular aspects of buildings, especially in circumstances where the detailing may be unfamiliar to the designer or to the design office. On large projects and projects with unusual details, many manufacturers will offer to provide the technical drawings and written specification clauses for the designers; for example, cladding companies will provide a complete package. This saves the design team a lot of production work, shifting their emphasis to coordination and checking information from other sources.

Representatives form an important link between manufacturers and potential specifiers of their products, but they are an expensive resource, and not all manufacturers employ them. They have a dual role, employed both to raise the awareness of the specifier to their employers' products (a marketing role) and to provide information and help to the specifier with the aim of getting the specification (a technical and sales role). For the purposes of this book, the term 'trade representative' is used to cover both sales representatives and technical representatives.

While trade literature tends to describe and illustrate typical solutions, in many circumstances, the specifier is faced with anything but typical situations. Thus, further information has to be requested from the manufacturer, and details may need to be discussed over the telephone or face to face with the manufacturer's trade representative. It is this situation that defines the specifier's requirements of a trade representative. However, like designers, the quality of trade representatives varies, from those with excellent technical knowledge of building and their organization's product range to those who have limited experience of building but are good at selling. Naturally, the former are very useful to designers, but the latter are regarded as a waste of time because they are unable to answer the technical questions asked of them (furthermore, many designers do not like being sold to).

The quality of the trade representative may be determined by the policy of their employers. Little research has been done on this, and the information would doubtless be considered commercially sensitive, but some manufacturers appear to concentrate on the selling function, while others attempt to provide good technical support. The common attitude within the design community to trade representatives is that they are regarded as a waste of time by offices and, if they must be seen, are given the cheapest person to talk to: a junior member of staff who will have little influence on specifying policies. This is either unknown to product manufacturers or ignored. The former is possibly because their advertising is designed by advertising agencies more familiar with other markets and unaware of the culture of design and engineering offices. This culture and the mechanisms by which offices handle trade representatives are dealt with below.

Specifiers require technical knowledge and technical information to be provided quickly and accurately, usually for a very specific purpose. This means that trade representatives who have empathy with the designer's concerns and can answer their queries quickly will be influential in helping the specifier to choose their products over those of a rival manufacturer. Another strategy employed by manufacturing organizations is to provide a technical helpline to answer technical queries. Sometimes, these are provided instead of trade representatives, sometimes in addition to them. Quick and accurate responses to specific technical questions will be expected by specifiers, and the speed and quality of the response may influence their decision to use or reject a particular product. In view of the fact that providing trade representatives is an expense for manufacturers, one would imagine that they would want them to be as cost-effective as possible. The clear lesson from the work that the authors have carried out is that designers and specifiers require both technical literature and technical advice, the former at the design stage so that they do not need to waste time with technical enquiries, the latter at the specifying stage and produced in response to demand.

For many specifiers, the service provided by the manufacturing company and/or supplier is equally as important as the characteristics of the product. Help with detailing difficult junctions and writing the specification will be welcomed by busy designers with tight deadlines. Technical helplines and the prompt visit of a trade representative to assist and provide product specific knowledge are important services that can give manufacturers competitive advantage over their immediate rivals. Typical services provided by manufacturers may include:

- guaranteed response to technical queries (within twenty four hours)
- bespoke design service and provision of free drawings, details, specifications and schedules
- provision of CAD files
- structural calculations for submission to building control
- on site technical support
- product specific guarantees and warranties
- access to accredited installers.

Efforts to develop a working relationship among manufacturer, designer and contractor are a small investment for all parties to ensure a relatively trouble-free 'partnership'. These help to ensure future specifications for the manufacturer, usually via proprietary specifications.

Touching the product

A service that manufacturers may provide is to supply samples of their products, usually through their representatives. By the time the specifier is requesting such a sample, a preliminary decision to use that particular product will already have been made. At this stage, the function of the sample is to confirm the choice rather than to select among the options available; something that occurs at an earlier stage in the design. What the designer needs then is a collection of samples from a range of competing manufacturers. A resource that has been available to specifiers in large cities has been the building information centres that provide a shop window for a variety of manufacturers to exhibit samples of their products and display their trade literature. They have always been popular with specifiers because they provide an opportunity to touch, assess and compare products without any pressure from the manufacturer's sales representative. However, the range of products on display is both limited by the size of the showroom and restricted to those manufacturers prepared to pay for the space. Thus, the products on display tend to be those of larger manufacturers with larger marketing budgets rather than their smaller, less affluent, competitors.

Naturally, the ability to look at and compare different materials is particularly important for those products that affect the finished appearance of the building. During the 1990s the brick manufacturer Ibstock established a series of design centres in major cities throughout the UK as a means of bringing their bricks to a larger audience (Cassell, 1990). Other commercial concerns also operate showrooms for particular product types. For facing brickwork, there are several brick showrooms around the country. Known as 'brick factors', they act as a middleman, stocking a wide variety of bricks from different manufacturers and aiming to encourage the designer's specification.

Manufacturers may also invite specifiers to their production facilities to see the product being manufactured and/or offer to take specifiers to construction sites and completed buildings to see their product in use. These visits serve a similar function in helping the specifier to see the product and, in the case of the production factory, also the quality control procedures. Such visits help the specifier to gain a thorough understanding of the product, while the manufacturer hopes that by taking the time to inform the specifier they will stand a good chance of being named in the written specification. Given that the specifier has invested some time in their product, he or she may be more reluctant to change the specification if requested to do so.

Another source of information are the centres that aim to promote environmentally friendly building products and systems. A well known example is the VIBA exhibition at s'Hertogenbosch in The Netherlands, which displays materials and arranges educational events to raise the awareness of specifiers to alternative ways of detailing and specifying buildings with the aim of minimizing the environmental impact on our planet.

Builders' merchants are another source of information about products stocked and available in a particular location. Although these depots may not be the designer's first port of call for information, for those keen to specify products readily available and clearly priced, they represent a good source. Again, the products are on display, although in less glamorous surroundings than the building information centres. For specifiers keen to use local suppliers and locally sourced materials to reduce the haulage (and associated pollution), the local builders' merchants provide a good source of information.

End of chapter exercises

- Take a material, such as concrete, steel or timber, and conduct a search for relevant information sources. Make a note of your sources and try to identify how independent or otherwise the information sources are.
- Based on your initial material choice, take an element (e.g. a concrete beam, steel lintel or timber beam) and conduct a search for manufacturers of these products. Based on the results of your search list three manufacturers that you would like to consider in further detail and explain why you have chosen these three.
- Now compare the three manufacturers and choose only one of them. Explain your decision.

4 The selection process

The development of a common market in the European Union and increased globalization of markets mean that specifiers have an enormous range of materials from which to choose. In the UK alone there are approximately 20 000 building product manufacturers, many of whom offer more than one product for sale (Edmonds, 1996). Manufacturers, in response to competition, new regulations and changes in architectural fashion, continually introduce new products. In addition to these new products, there are numerous minor product improvements that are constantly introduced by manufacturers to prolong their products' life on the market. These new products and product improvements, like the established products, are dependent on decision makers in the building industry for their selection. In his book *The Roots of Architectural Invention*, Leatherbarrow (1993: 143) made the observation that:

> Because materials are familiar in experience and unavoidable in construction one might assume this specification is a procedure that can be described simply and clearly; in fact the opposite is true, for it is both a rarely discussed procedure and one that exposes strikingly obscure and indefinite thinking when questioned. Yet this obscurity is unavoidable because material selection is inevitable.

Product selection, above all else, is one of the most important considerations for the long term durability of the completed building and an area in which building designers should excel. Not only have the selection and specification of building products attracted little attention from researchers, but they are rarely discussed by practitioners, presumably because it is difficult to separate the designer's goals from building materials as entities in their own right (Patterson, 1994). It is a process that is often seen as unglamorous and something that should be carried out as quickly as possible. Architectural literature, especially architectural periodicals, tends to be preoccupied with the finished appearance of buildings, both inside and out, paying little attention to the process of producing the building or to its long term performance. The majority of published literature that has investigated the way in which architects make decisions has concentrated on the design process with emphasis on the resulting form (Rowe, 1987).

This body of literature, going back to the 1960s, is commonly referred to as 'Design Methods' literature and is primarily concerned with creative problem-solving (e.g. Thornley, 1963; Heath, 1984). There is a clear distinction in this work between the selection of materials, which may be an inherent part of the designer's design idea, and the selection of specific products to fulfil this vision. The separation of stages can also be seen in some of the larger architectural offices in which the design architect delegates responsibility for the detail design to an architectural technologist or technician. However, there is evidence that the separation of conceptual from detail design, which has been effectively institutionalized in France with the different types of work carried out by different offices, is not liked by practitioners in Britain, a point noted by Mackinder (1980: 12):

> Many architects assert that [the design process] 'is' the process of selection, organization and specification of materials, and refuse to separate the two.

The irony is presumably unconscious with an apparent difference between the architect's perception of the process and the actual process, although practices differ in different offices.

Various, and often conflicting, selection criteria that may be employed by specifiers to achieve their objectives have a major influence on what is eventually confirmed in the written specification. The specifier's office does not work in isolation, it is engaged to provide a service to a client (who may contribute to the process of product selection) and it is involved in communicating the design to the contractor on the building site. Thus, the specifier, even in the most simple of contractual relationships, may be influenced by contributions from persons outside the design office. The various people involved in the process, each with different values and goals, are brought together for one particular project, during which both formal and informal communication will take place between them. The manner in which the various individuals interact will depend to a large extent on the nature and size of the building project and the type of contract used to procure the building, which can vary widely among the projects being worked on by the office. In some projects the participants may work very closely as part of an integrated team, while in others there may be more space between the participants and less frequent communication (Emmitt and Gorse, 2007). So the specifier's choice is influenced not only by technical considerations, but also by factors covered under the umbrella of design management.

Contributors to the specification process

Until quite recently it was the architects and engineers who took the vast majority of specification decisions, supported in the production of the written specification by a small army of technicians and assistants. With the introduction of new procurement methods, increasing use of off site manufacturing and the growth of new specialists such as architectural technologists, a wider range of actors is now involved in the specification process, including the client and building users.

Control of the specification may be by the designer and engineer in traditional contractual relationships, but is more likely to rest with the contractor on design and build type arrangements. Regardless of this, the specification decisions are usually made within the environment of the design or engineering office (which may be in house in contractor led relationships). There are several factors that are under the control (to a lesser or greater extent) of people or organizations outside the specifier's office that will influence the specification process. Factors under the control of the manufacturer are the cost of using the product, the availability for delivery to site on programme, the associated services provided and the quality of the product information. Other factors are produced by those who may be regarded as members of the temporary project team. These include the client's preferences, the influence of the town planner and limitations imposed by the main contractor.

Mackinder (1980) listed a number of factors that influenced the architect's selection of building products, which she called 'external' influences (from the perspective of the office). These were the client, the quantity surveyor (QS), the contractor, the government, and the role of structural, mechanical and electrical consultants. Since Mackinder published her research there has been a move to more cohesive and collaborative teams and some of these influences would now be considered to be an integral part of the design and construction project team. Emmitt (1997) found that the various actors described below tend to exert different pressures at different stages in the specification process, which were termed constant variables on projects. The Barbour Index has investigated the role of different actors in the specification process on an annual basis since 1993. Some of these reports have focused on the role of specific actors, such as clients (1995, 2003) and contractors (1994, 2004), while the 1993 Report *The Changing Face of Specification* and the 2000 Report *Influencing Product Decisions: Specification and Beyond* provide a more comprehensive insight into the factors that influence product selection and specification. The extent to which these actors influence the specifier's decision making process is explored and illustrated in the case studies reported later.

The client

The client (the building sponsor) will determine the overall quality of the building by both the amount of money and the amount of time available, thus defining the limits within which the design team has to work (sometimes exceeded).

Mackinder's (1980) work found that certain clients influence design decisions and hence product selection more directly, mainly through the use of a client's standard specifications, although four of the architects that she interviewed believed that the client should not be involved in the specification process. There is doubtless some truth in the fictional account in Creswell's *Honeywood Settlement*. The client, Sir Leslie Brash, has a conversation in his club and is persuaded of the merits of a brand of paint, which he instructs the architect to specify. The result is a spectacular failure when the paint falls off the walls.

Emmitt (1997) also found evidence of client involvement in the choice of product. The Barbour Index (1993, 1995, 2000, 2003) has indicated that the client has a major role to play in the selection of products, although in the 1993 report the clients said that they would rather not be involved if they could avoid it. The Barbour Reports published in 1995 and 2003 looked at the influence of 'major clients', such as house builders, the National Health Service (NHS) and local government departments. These reports demonstrated a trend towards greater involvement by clients in the specification process. Although there was a wide variation between different sectors, the 2003 Barbour Report showed that 31 per cent of major clients were involved in making brand (proprietary) decisions for new build and 42 per cent for refurbishment projects. This represents an increase on the 1995 and 1993 figures. It is not possible to say why this trend is developing, other than to note the greater involvement of the client in the design process, as promoted in the UK by the Commission for Architecture and the Built Environment (CABE) in an attempt to improve both building quality and the value provided to the client.

In the 2003 Barbour Report clients reported a preference for traditional forms of contract because they retained control over the products used and hence the quality of their buildings. The clients also noted the importance of team assembly because it has an influence on the product decisions, with almost half of the sample appointing designers known to them (from approved or preferred lists of companies). Relationships with manufacturers appeared to be related to the size of the construction budget; the larger the budget the more likely the client had a close relationship with manufacturers. Partnering agreements existed between clients and, for example, suppliers of kitchens, tiles and paint, to ensure quality and achieve price discounts. Eleven per cent of clients said that the relationship between them and the manufacturers could be improved and furthermore that products could also be improved with regard to sustainability and life cycle costs.

One third of clients included product references in their initial brief, which was especially prevalent with the house builders and the retail and leisure sectors. Briefs for refurbishment projects were more likely to contain references to proprietary products compared to new build schemes. Most brand specifications originated from approved product lists. Clients with approved product lists increased from 46 per cent in 1995 to 56 per cent in 2003. Although the approved lists are highly influential, only 38 per cent of clients claimed to have a formal review process for assessing products. Instead, the approved lists were developed as a result of good and bad experience of specific manufacturers and products. It is also significant that the clients said that 42 per cent of their brand specifications were challenged by the architect (20 per cent) and the contractor (22 per cent), although they also said that they did not always agree with the alternatives suggested.

It should also be mentioned that there has been a move towards greater involvement of the building users in the design process, and this can have an effect on the choice of certain building products. For example, some social housing schemes encourage potential tenants to choose their internal fittings, such as kitchen units and the colour of the bathroom suite, from a predetermined range (usually provided by one manufacturer or supplier). Users and user representatives may also contribute to the briefing process and the development of the client specification in some projects. The authors were, however, unable to find any research into how this influences the specification process.

Consultants

At its most simple, the design team usually includes a structural engineer and a QS in addition to the architect, and is influenced by the client's budget and complexity of the project. Generally speaking, the larger the project, the greater the number of different consultants who may, to lesser or greater extents, contribute to the individual specifier's decision making process. It is important that such consultants be kept informed of decisions that may affect their work. For example, if changes to finishing materials mean changes in the weights of those materials then the structural engineer needs to be informed. Such a change may have an insignificant effect on the structure of a floor, but may be quite significant in a roof.

Architects and engineers also design bespoke elements for buildings that are not available from manufacturers' standard ranges. This is often done in close collaboration with specialist suppliers and craftspeople, resulting in the use of prescriptive and performance specifying (and the use of nominated suppliers in the contract documents).

Quantity surveyors

In Britain it is not uncommon for the QS to contribute to the preparation of the written specification. Many QSs complain that they are forced to provide an input to the written specification because the document is sent to them containing errors and omissions. This may be considered as careless or just lazy practice on the designer's behalf, although the situation usually arises because insufficient time has been allocated to the task, i.e. it is a management problem.

Direct economic advantages or economic disadvantages to the client are different from those of the specifier, who may not always be aware of individual product costs. The QS offers cost advice to both the client and the architect. Because of this, it may be that the specifier is not always aware of the cost of individual building products. Instead, it will be the QS who has access to such information, both from information provided by product manufacturers and from experience from previous projects (indeed, it may be the QS who will suggest alternative, cheaper, products to the specifier). Three quarters of Mackinder's (1980) sample of architects said that they left some items to be specified by the QS, usually because they were short of time. However, the QS was seen by the majority of the architects interviewed as a 'cost information file' and was asked to suggest alternative solutions when cost was very restricted. Emmitt (1997) also found evidence that the QS influenced specification decisions, but less than that reported by Mackinder. The Barbour Reports indicate a trend towards greater involvement in specification decisions by QSs between the 1993 and 2004 Reports. This may suggest a change in practice, although the methods of data collection were different, which makes comparisons difficult and potentially misleading.

Town planners and building control officers

The proposed building design must have planning permission and Building Regulation approval before it can be constructed. Legislative control may influence the selection of materials, especially materials that form the external fabric of the building. The architects interviewed by Mackinder (1980) said that the Building Regulations affected their selection of materials where fire prevention and thermal insulation was concerned. Architects also said that the planning officers exerted a 'very strong' influence on the external appearance of buildings and hence the choice of products, in three quarters of her sample. Planners were also described as often being in the habit of stipulating proprietary products in order to achieve a particular effect known to them; examples quoted of this type were *Velux* rooflight windows and *Dorking Brick* (Mackinder, 1980); a trait also revealed by Emmitt (1997). The town planner's possible contribution to the specification process where products are visible on the exterior of a building presents clear dangers as the planner has no responsibility for the building's performance.

Building control officers may also influence the specification process. For example, Chick and Micklethwaite (2004) noted that in their interview sample a building control officer perceived new materials as a risk. Where building control officers are not familiar with a product, it may be necessary for the specifier to supply the necessary information.

The prime contractor

Since Mackinder's research was published the role of the contractor has changed, with many contractors taking greater responsibility for design and specification decisions and/or becoming involved with the design team earlier in the design process. The design team increasingly uses contracts and specifications that place a design responsibility on the prime contractor and the sub contractors. This shifts the act of selecting products to the contractor. The reasoning behind this development is that the contractor is better placed to select products which offer better value to the client and it is therefore common to use relatively open specifications when building to a fixed cost. The implications of this in terms of future maintenance and sustainable credentials need to be investigated. One might imagine the contractor to have little interest in this aspect of a product's performance, unless the contract involves the operation as well as the design and construction of the project.

The Barbour Index has published two reports that have looked at the contractors' influence on the selection process, published in 1994 and 2004. From the contractor and sub contractors' perspective the most important factor influencing product decisions is the cost of products. This was identified in the 1994 report and remains the most important factor, despite the fact that health and safety, life cycle costs and sustainability issues are becoming more important. These reports have focused on major contractors and specialist sub contractors with a large financial turnover. Design and build contracts account for approximately 75 per cent of their business and as a result they are heavily involved in product decisions. In the 2000 Barbour Report the main contractors estimated that they selected approximately one quarter of all products. In the 2004 Barbour Report the contractors claimed to select approximately 50 per cent of all products, reflecting an increase in contractor led contracts, such as design and build and private finance initiatives (PFIs). In these types of contract the contractor typically issues a set of product guidelines to be followed by the design team and over one third of contractors also issue an approved product list.

The main contractor as actual purchaser of the building products has a different relationship with the building product manufacturers than the specifier. Very large contractors claim to have direct relationships with preferred manufacturers and some of the smaller contractors expressed a desire to have better working relationships with manufacturers (Barbour Index, 2004). However, most contractors and sub contractors have accounts with one or two builders' merchants, who themselves will stock a range of popular building products, and by placing their orders with a particular merchant they can obtain some financial discount. Most large to medium sized contractors have a central purchasing department (buyers' department) to order materials in bulk, thus the site manager's relationship with the manufacturers is via the central purchasing department, unless there are problems (in which case the relationship is with a technical representative). Some smaller contractors will order materials as required, in response to requests directly from the site. Problems tend to arise when the specifier has specified proprietary products that are not stocked by the builder's normal merchant, in which case, the contractor may request a change of product to one that is familiar to him and is available from his regular merchant, but which is not necessarily familiar to the specifier. This is confirmed in the Barbour Reports, with contractors trying to change 23 per cent of product decisions and claiming to be successful in 80 per cent of cases (Barbour Index, 2004). Emmitt (1997) also found evidence of pressure by contractors to change products, and the implications are explored further in Chapter 6 and illustrated in the case studies.

Sub contractors and specialist suppliers

The Barbour Report 2004 looked at the growing influence of specialist sub contractors in the selection of products. The sub contractors claimed to be consulted frequently by the main contractor over which products should be used on projects, although this tended to be on design and build contracts. On these contracts it was common for approximately 50 per cent of products to have been specified before the sub contractors became involved, with the remainder decided after they had been appointed (since many sub contractors also supply design work). Sub contractors and specialist suppliers have also claimed to change specified products, often without the knowledge of the prime contractor (Barbour Index, 1993), the implications of which are again explored in Chapter 6.

Detail design decision making

> ... the selection of mutually compatible components to solve some specific defined standard problem or some specific unique problem within a larger consistent design context (Wade, 1977: 281).

At first glance, the selection of materials and products to meet a specific purpose would appear to be a relatively straightforward activity. However, on closer inspection, both the issues to be considered and the actual process are complex, and one of the central problems facing designers is that of determining priorities. Resources are not infinite, so to achieve the given objectives within the constraints of time, finance and expertise, the number of variables considered has to be limited, i.e. designers must determine their priorities for each design project. Furthermore, the selection of one product over another will affect the selection of other components because no architectural component can be considered in isolation: the designer must constantly appraise and reappraise the product in relation to the building assemblage as a whole. Detailing a building is a process of continual evolution grounded in what the designer (and the design office) believes to be best practice. Detailing is based on combining information obtained from a wide variety of sources and influenced by many different contributors. It is also about the choice of the correct solution for a particular set of circumstances at a particular time, considering the benefits for clients, builders and users within a framework of limited resources and creative endeavour.

Underlying all issues concerned with design, manufacture and assembly is the ability to make decisions in the available time. Regardless of the building type and size, and the complexity of the design, each project will have some form of time constraint imposed on it. Usually, the client requires a completed building for a particular date, a date that will influence the amount of time allocated to different phases of the project. This imposes time constraints that have to be accommodated into overall programming of resources, thus limiting the amount of time available for producing the requisite information. Adequate time is required to consider appropriate products and set performance standards, coordinate information provided by others, write the specification and check the project documentation for consistency and errors. Unfortunately, many clients are unaware that a shortage of time will have an adverse effect on either the quality or the cost of the finished product, and sometimes both, especially when the task of writing the specification is given to those not fully briefed in the design requirements. While almost any design office would be able to provide anecdotal evidence of this it would be difficult to collect evidence on the full effect of time constraints. Design is a particular type of decision making activity, and one about which there has been much research and debate, although whether or not there is yet an adequate 'natural history' of the design process is debatable (Yeomans, 1982), but Rowe (1987) provides useful, and critical, insights into design thinking.

Programming of the specification process is paramount if good quality information is to be produced. Time constraints also influence the uptake of new products, as demonstrated by Mackinder (1980) and Emmitt (1997). When a project had to be completed quickly, there was an increasing tendency to stick to products used on previous projects, thus eliminating the time needed to search for alternatives. The extent to which this action hinders the uptake of new products is explored in later chapters.

A slightly different perspective is that design is an information processing activity (Akin, 1986). This, and associated work, is based on the premise that designers, managers and hence organizations can be understood by observing their decision making behaviour and then designing and implementing an information processing model (Simon, 1969; Newell and Simon, 1972). Heath (1984) concluded that the appropriate method is determined by the social nature of the task, so that different methods need to be used to suit each particular situation. They claim that through the use of these information processing models, designers can select from a range of tactics to shorten the time required to produce the design and to reduce the potential for error. This observation is relevant to the conceptual and detail design stages.

Decision making

Much of the literature dedicated to design decision making is centred on the actions of individuals, with less emphasis placed on the collective efforts of design organizations and the building 'team'. Clearly, it is easier to observe the behaviour of individuals, especially in controlled experiments, rather than the group activity of a design office. Therefore, this bias in the data available is to be expected. In practice, however, individuals are constrained and influenced by the behaviour of the group to which they belong and by other groups party to the design project. There are also cultural constraints. When people make decisions, they tend to follow rules and/or procedures that they see as appropriate to the situation (March, 1994). This is particularly so of professionals who are expected to act in a manner appropriate to their particular professional background. Designers not only have to satisfy their client, but also have to satisfy different building users and, as in many other professions, will feel a need for peer approval.

Because designers will face design problems that are ill defined, poorly described or diffuse in nature, attempts must be made to define the problem clearly before it can be resolved. In some of the design literature, this process is described as 'questioning' (e.g. Potter, 1989). Definition of problems is made easier through the designer asking questions of himself or herself, and also of others. The aim of this questioning process is to be able to take full account of the information, explore possibilities and recognize the limitations, essentially a process of simplification.

Building designers are expected to act in a logical manner when selecting building products and materials, assessing all the options, against a background of legislation, before making a choice. Nevertheless, research suggests that this may not be the case. Although much of the literature on decision making makes assumptions based on rationality, the validity of this assumption is thrown into question by studies carried out by behavioural scientists. Observational studies of decision making behaviour suggest that individuals are not aware of all the options, do not consider all of the consequences, and do not evoke all of their preferences at the same time. Rather, they consider only a few options and look at them sequentially, often ignoring some of the available information (March, 1994). Decision makers are also constrained by incomplete information and their own cognitive limitations. Thus, although decision makers may set out to make rational decisions, in reality, they make decisions based on limited rationality: they search for a solution that is 'good enough', not the 'best possible' solution.

The factors that place constraints on human decision making are:

- *Attention span*: it is impossible to deal with everything at once. There are too many messages, and too many things to think about. Thus, we tend to limit our attention to one task at a time, ignoring messages that are irrelevant to that particular task, engaging our selective exposure. Our attention span is also limited by time.
- *Memory*: both individuals and organizations have limited memories. Memories are not always accurate; we tend to remember facts as we like to see them, rather than as they actually happened. Organizations and individuals are limited by their ability to retrieve information that has been stored. Records are not kept, are inaccurate or are lost, so that lessons learned from previous experience are not reliably retrieved. Moreover, knowledge stored in one part of an organization cannot readily be used by another part of that organization.
- *Comprehension*: despite having all the facts to hand, the relevance of information may not be fully understood. There can be a failure to connect different parts of information. Furthermore, individuals have different levels of comprehension, making it difficult to foresee how each will respond to the information they have. For example, the architect, manufacturer and contractor may understand the same piece of information differently simply because of their different backgrounds.
- *Communication problems*: specialization, fragmentation and differentiation of labour encourage barriers and present difficulties in the transmission of information and knowledge. Different groups develop their own frameworks and language for handling problems, and communication between these cultures can become difficult.

One way of reducing the effect of these difficulties, while reducing the time required to make decisions, is to use familiar solutions. Designers draw on experience (their own and that of others) to come up with a particular design solution for a specific site and a particular client. Relying on personal experience requires a considerable amount of knowledge of various solutions to problems, knowledge that is only acquired with experience. Young designers (with limited knowledge) will, to greater or lesser extents, rely on solutions suggested by others, notably their more experienced colleagues within the office. Solutions provided by office standard details and standard specifications represent the expert knowledge of the organization that produced them. They are also the result of the careful selection and use of pertinent and current information, having the effect of limiting the amount of information available to the designer. However this can only be done by sifting information as it arrives and rejecting a large proportion, which implies some constantly applied management of information within the office (Emmitt, 2001), as discussed in Chapter 3. Investment in this kind of information management is expected to reap benefits in simplifying the decision making process.

Mackinder (1980) observed that the way in which architects make decisions is closely influenced by two factors: the amount of time available for the scheme to be fully produced (the most critical factor) and the overall importance of the material or product in relation to the total project. In addition, when an architect's office was involved in projects where speed was of prime importance (industrial or commercial projects), there was a strong tendency to adhere to a well established vocabulary of materials for structural forms, cladding materials and finishes. The more individual the project, the greater the amount of time was said to be required for research and development work. Mackinder concluded that projects involving any extensive use of new materials or systems (no previous knowledge or expertise) can only be entertained if time is available.

Personal collections of literature

Research into specifiers' behaviour has consistently found that the majority of specifiers retain a personal collection of trade literature, independent of the office library. Many specifiers have their own favourite products and sources of information based on their individual experience. This is represented by a 'palette of favourite products', from which they always choose unless forced to do otherwise (Mackinder, 1980; Emmitt, 1997). Evidence suggests that the other players in the network (e.g. town planners) also have their own palette of favourite products, which, for various reasons, they will want to use on a project. This may be financially motivated in the case of the contractor and personally motivated in the case of the town planner. Use of products from this personal collection reduces the amount of time spent searching for products to suit a particular situation, and because they are known to perform well (or more to the point known not to fail), their selection poses little risk to the specifier. This palette of product information is essentially a knowledge base that is used to aid the specification process. It may be maintained as a file of paper literature or in a digital file that can be quickly imported into drawings and specifications. This is an important information source and is investigated further in the case studies below (see Chapters 9 and 10), but the advantages and disadvantages of the use of personal information within the design office are that:

- it saves time looking for information
- the product is familiar so that uncertainty is reduced
- details and specification clauses can be imported from a previous job.

In contrast, the disadvantages are that:

- it reduces the likelihood of specifiers looking for alternatives and so may hinder innovation
- there is a greater chance of superseded information being used
- there is a greater chance of error through the use of rolling specifications.

The collection of literature makes it quick for the specifier to check details about products that are already known to them and their office. However, quality management systems prohibit the use of individual files of product information because of the danger that they may be out of date. Nevertheless, research by the authors found that specifiers went to great lengths to maintain their own file of information in spite of such prohibitions. With the increased use of computers, the palette of favourite products can be easily stored in an electronic file and quality managers need to ensure regular audits either to eradicate the use of personal sources or to ensure that the information contained in personalized files is current and in accordance with office standards. Eradicating personal sources of information may be the easier of the two to manage.

Similarly, the majority of design offices build up some experience of successful and unsuccessful products and details over time. Wade (1977: 158) has commented on the reliance of known sources:

> Each office develops its own library of product information and begins to develop preferences for the use of some products and not for others, as a result of good and bad experiences of those products.

This experience may be disseminated through internal memoranda or office standards, leading to the development of an office palette of favoured products, possibly incorporated into standard details and the master specification. Such practice reinforces established patterns of behaviour and discourages independent thought or action, confining specification to products from the list of approved products and preventing the use of 'blacklisted' products, essentially forming a further barrier to the use of innovative products.

Lists of approved products

Most design offices maintain a list of approved products, essentially an office palette of favoured products. Products that are known to perform, through both care in their selection and subsequent experience in their use, are included in the approved list. Such lists help the organization to ensure quality and compliance with particular standards and regulations. They also help to reduce exposure to risk (as perceived by the owners of the business) and help to save time for the specifier searching for information. As noted above, some clients insist that the design team use certain products or suppliers because they have good experience of using them. In contrast, some products and/or manufacturers may be included on a prohibited list because of previous bad experience. Inclusion on the list may be because the product failed in use, was found to be of poor quality when delivered to site, or was difficult to use on site, or simply because the service provided by the manufacturer was found to be unacceptable. Poor service by the manufacturer may include a failure to respond to technical queries on site or failure to deliver products to site as scheduled. Care should be exercised in the use of lists of both approved and especially prohibited products, since manufacturers may be less than pleased to find themselves excluded from the former or included in the latter.

Product selection criteria: fitness for purpose

Regardless of where a specifier happens to work, be it in the office of the designer and engineer or in that of the contractor, he or she will need to select materials, components and products to satisfy the design intent. Information provided by manufacturers, sometimes supplemented by verbal information from the trade representative, should enable specifiers to make an informed choice about which particular product to select. These products must be 'fit for purpose', i.e. they must suit the particular requirements of the project and also comply with prevailing legislative restraints.

Although specifiers are bound by professional ethics that prevent them from accepting any financial inducement for selecting a particular product, they are unique as consumers because they are selecting products that are themselves products of a design process, marketed through carefully designed advertising and technical literature. Therefore, there is the possibility that specifiers are more likely to select products that they can empathize with, i.e. products perceived as being most sympathetic to their design values (e.g. Grant and Fox, 1992). They may be more receptive to advertising literature that accords with their own design philosophy in terms of style, composition and colour, etc. Thus, competing products may be analysed subjectively rather than objectively. In addition, there is a difference between the goals of designers and those of contractors. Contractors will be more interested in availability, buildability and profit margins; an important point to bear in mind when using performance specifications.

When designers are asked what criteria they employ, they are most likely to say first that they are looking for products that are fit for purpose, i.e. that suit their particular requirements. Fitness for purpose can be broken down into a number of interrelated criteria for product selection. Some of the factors described below, such as durability and cost, which might be grouped under the word 'quality', are of direct concern to the client. Some others that are now beginning to influence choice are the result of legislation. Such factors include the safety of the product (both during construction and in use), its embodied energy and its environmental impact. In practice, some of these factors may well be in conflict and specifiers have to resolve these conflicts within the time and budget available using their professional judgement. Some of the most common factors are described in more detail below, listed in alphabetical order (the list is not meant to be exhaustive).

Appearance

For designers, the aesthetic appeal of a product is often at the top of the selection criteria, especially where it is to be seen and experienced when the building is complete. Thus, for the designer this is often the first filter that is applied in the selection of materials and products before any detailed analysis of their performance characteristics. The Barbour Reports have consistently found that specifiers prefer to specify proprietary products if they will be seen within the completed assembly and are more likely to resist substitution requests. Aesthetics is very much a personal matter, with different specifiers having their favourite materials and manufacturers, but clearly whatever is used must be appropriate to the overall aesthetic of the scheme. Furthermore, many design offices develop a particular architectural style that may rely on the use of certain materials as an underlying approach to all their building projects. When dealing with an existing building, aesthetics takes on a different slant, especially when trying to match new materials to existing materials. Here, the specifier is constrained by the existing fabric, and so choice will be constrained in a different way.

Availability

Availability can have major implications for the programming of the construction works, both for the designer and for the contractor. Checking availability at the specification stage can help to eliminate problems later in the contract.

There is no point in specifying something that cannot be delivered in time for construction. This is true of both proprietary and performance specifications; there is little point in writing a specification for a particular element that takes six months to manufacture and deliver to site when dealing with a fast track project unless it can be preordered and programmed for. Availability should be checked directly with the manufacturer. This is especially important when materials, components or systems are being specified that take time to transport to the site and/or have to be manufactured to order. Early involvement of project managers, specialist sub contractors and suppliers will help to define clear and achievable programme deadlines. This can be of great assistance to the specification writer in helping to determine his or her choice of product(s). Note that availability, or rather the lack of it, is often used as an excuse by contractors for substituting products on site, as illustrated in the case studies.

Constructability

Products cannot be selected in isolation from their neighbours in the overall building construction. The manner in which the building is assembled, its ease of constructability (buildability), is an equally important consideration for the design team and those responsible for the building's assembly. The integration of a wide range of technologies, comprising the structure, fabric, services and finishes/installations, in a safe and efficient manner is a prime concern for all members of the project team. Of particular concern to the prime contractor and sub contractors is the ease of handling of products on site. Architectural detailing and the specification must consider the requirements of those charged with constructing the building and those charged with maintaining it.

Design criteria

The physical characteristics of the building site, requirements of the client and approach of the design team conspire to set the design agenda. The resultant design criteria establish the critical test for all the products specified.

Durability and maintenance

We cannot aspire to the theoretical perfection of Oliver Wendell Homes' wonderful one hoss shay (the Deacon's Masterpiece). Tired of individual components wearing out at different times, the Deacon's Masterpiece was designed so that all the components lasted for the same length of time, until it all fell to pieces in a moment. In contrast, the designer should be concerned that the life of individual components is commensurate with either the design life of the building or the life of its anticipated replacement. Some components may have a life similar to that of the building itself, while others may be designed for regular replacement. Moreover, the design life of a component may also be related to maintenance costs. The life of timber windows and doors can be measured in hundreds of years, as demonstrated by the survival of eighteenth century joinery to the present day. However, such joinery has not survived without frequent repainting. Modern plastic based window frames do not need this regular maintenance, but have a limited life, so that their long term cost may well be greater than that of an equivalent timber frame. Here the balance between the cheaper initial cost of the plastic frame, against the long term saving possible with the timber frame, is the kind of issue that might need to be discussed with the client before a choice can be made.

Internal and external finishes are dependent on the materials used and the manner in which they stand up to weathering and daily use. Mechanical and electrical components will wear and need maintenance, but some components may need to be replaced while still perfectly sound, simply because of higher expectations. This may merely be a matter of fashion. At the time of writing, Belfast sinks are being sold for up market kitchens while those removed some years ago to be replaced by stainless steel are having a second life as garden planters. More seriously, hotels may feel the need to upgrade bathrooms simply to meet customers' increasing expectations.

The physical characteristics of components will determine the durability and performance of the building. As indicated above, the anticipated life of building products needs to be considered at the specification stage. Information on durability is usually provided by the product supplier requiring the specifier to place a certain amount of trust in that information. However, durability may also be built into the common manufacturing standards and enshrined in national standards such as British (BS), German (DIN) or American (ASTM) standards. Although these may not actually define the life of the product, they may require properties that will ensure a satisfactory life, especially where this affects public safety. Consider, for example, flue liners that are attacked by flue gases and where failure allowing their escape would be hazardous. The durability of these may be indirectly controlled by specifying a minimum thickness. However, such standards are sometimes questionable. An example is provided by steel cavity ties that rely upon the thickness of the galvanizing for their durability. This was fixed by a British Standard, which was at one time changed to allow thinner coatings. The result was a rash of failures when the thinner coating proved inadequate and the coating's thickness had to be increased.

Frequency and ease of replacement are considerations for items that have a limited service life, as should be ease and cost of maintenance when selecting one product over another. It is becoming increasingly common for building-insurance organizations to lay down strict criteria that can influence choice of products, especially where they are concerned with asset and facilities management. Materials and components with short service lives or those that carry a greater degree of risk of failure may effectively be blacklisted.

Ethical and altruistic concerns

Some designers and design organizations hold very strong views on particular issues that inform their design thinking and decision making, the most obvious being a concern for the natural environment. Other 'political' factors concern the ethical practices of those in the supply chain (for example, how employers treat their workforce and the manner in which products are resourced and processed). Sometimes these factors are stated explicitly by the design office and/or its members, although often these issues tend to emerge during the life of a project.

Environmental credentials

The building sector is a major consumer of raw materials and energy, and with growing concern about the impact of the built environment on our natural world has come greater attention to the materials and products specified, with a focus on 'green' buildings and building products. It is difficult to find clear definitions of what actually constitutes a green building, or indeed a green product. The word 'green' tends to be used as an adjective to mean environmentally friendly. High performance building is another term that is sometimes used to describe a building with better environmental performance than a more conventional building. Spiegel and Meadows (2006) acknowledge that perceptions and practices vary widely. They also sound a note of caution in that some products are now being marketed as green, although their technical specification may not have changed, a phenomenon known as 'greenwash', and something to which specifiers need to be alert. The environmental credentials of specific products are not always easy to quantify in absolute terms and this can result in an element of subjectivity being introduced into many decisions concerning green products. The difficulty is that most specifiers have to rely on the information provided by the manufacturers, although there is a small amount of work that aims to provide guidance on the environmental impact of generic materials.

Reasons for specifying green (or comparatively green) materials and products vary, from response to voluntary and mandatory environmental guidelines and regulations to simple altruism. Although green buildings feature strongly in the architectural periodicals they do not constitute the bulk of buildings constructed in the UK. The majority of buildings follow the minimum requirements to achieve building control approval. The general perception is that 'green' represents something different from the familiar approaches and solutions; hence, the selection of green products represents a change in existing behaviour and higher perceived risk for the majority of designers and contractors.

Research based on a survey sample of 539 architects and 142 designers located in the UK, supported with follow up interviews with twenty of the respondents, found a raft of obstacles to the specification of recycled (green) products and materials (Chick and Micklethwaite, 2004). Obstacles were (in order of importance): lack of information, unfamiliarity, supply issues, cost, quality issues and practice constraints (time). They found that architects liked to 'play safe' and only specify materials with which they were familiar. Contractors had a very negative view, claiming that unfamiliar materials would raise costs and the building inspector felt that new materials posed a threat. Although Chick and Micklethwaite fail to acknowledge the earlier work on specifiers' behaviour, their findings support the work of Mackinder (1980) and Emmitt (1997), but contradict the more recent Barbour Report (2004), in which contractors expressed a more positive attitude towards green products.

Spiegel and Meadows (2006: 22) describe green products as being '... green in the way they are manufactured, the way they are used, and the way in which they are reclaimed after use'. This implies that the specifier needs to source information about the product and the manufacturing processes. Some manufacturers will comply with ISO 14001 or may use an eco management and audit scheme (EMAS), whereas others may adopt certain criteria, such as the use of recycled material in their product or using sources local to the centre of production. Anderson et al. (2002) recommend that specifiers ask manufacturers for Environmental Declarations, thus helping to provide some reassurance that the manufacturer is taking a responsible approach to the environmental performance of its processes and products. It is, however, very difficult for the specifier to determine absolute values and hence advantages of products and materials that are promoted as being green. At best, the advantages associated with green products are relative and contain some degree of subjectivity.

Experience and the lessons of failure

Research by Mackinder, Emmitt and the Barbour Index has shown that the familiarity with a product and manufacturer, from previous projects, is a major factor in product decisions. Good experience will usually result in the specifier using the same product and manufacturer on new projects. Conversely, bad experience may result in a product or manufacturer being ignored in the future.

All built structures will fail eventually, some sooner than others, although the end of a building's design life can be anticipated, and often extended, with sensitive and regular maintenance. It is the unexpected failures that cause the most concern for both designers (who may be liable) and owners because of the disruption and cost of putting it right. Failures of materials and/or execution can never be entirely eliminated, even with mass production techniques and their associated quality control procedures. Designers need to understand the limitations of the materials and manufactured products that they detail and specify, especially any changes in physical size due to moisture or temperature changes They also need to be aware of the practicalities of working and/or fixing the materials on site if failures are to be minimized.

Guarantees and warranties

One way of limiting the specifier's exposure to risk is to specify only products with written guarantees and warranties. Most manufacturers are happy to provide these for their products. To a certain extent the specification of products that have been certified by independent bodies, such as the British Standards Institute (BSI) Kitemark, gives the specifier some reassurance that the product will not fail, if used in accordance with the manufacturer's instructions. However, many products do not carry such certification and so it is common for the specifier to ask the manufacturer for written guarantees and warranties for their products.

Guarantees and warranties offer a degree of comfort to the specifier, although such insurances are only valid while the company is still trading, and only if the product has been installed as stated. Some products should only be installed by a number of specially trained and certified installers, which should help to ensure quality workmanship. There may be an increase in cost for this, but experience shows that the finished work is of a higher standard, and fewer problems occur during the installation. Some product specific guarantees may be valid only if the product has been installed by an accredited installer.

Initial cost

It is not an easy task to establish the cost of some proprietary products, nor is it easy to separate out the price of the product from the cost of the service provided by the manufacturer. This is an important point to make, simply because the service provided to the specifier will be different from that provided to the contractor. For example, the specifier will require technical details and possibly help with detailing and writing product specific specification clauses, and the contractor will require information on delivery, costs and possibly assistance with issues relating to buildability on site.

There is, of course, no such thing as a 'free lunch'. If drawings and specifications are provided free of charge by manufacturers (as discussed in Chapter 3), then the cost of this service is built into the price of the individual product. Companies offering such a service provide added value for specifiers, although one from which the client does not necessarily benefit.

The initial cost of proprietary products is not always known to the specifier at the time of selection. There are several reasons for this. First, manufacturers are often reluctant to give the price of their products to the design team, for fear that the specifier will choose simply on price and not on value. Second, in the UK and commonwealth countries, where it is common to employ a QS, it is the QS who obtains the cost information, not necessarily the specifier. The effect of this is that some designers show a lack of interest in the cost of building components. The price of the product to the main contractor will be determined by the relationship that the contractor has with the supplier and/or builders' merchants and the level of discount provided by the merchant on certain materials.

Cost reductions may be possible by selecting less expensive building materials and reducing the amount of time required to assemble them on site, but this assumes that these costs can be discovered.

Life cycle cost (cost in use)

It is not sufficient to rely only on the initial cost of the product as a selection criterion (although many specifiers and contractors do). The term life cycle cost covers the entire cost associated with the product over its life; this includes the initial purchase and installation costs as well as the cost of maintenance, replacement and eventual recycling. Otherwise known as the running cost or operating cost, the cost in use is set by the decisions made at the briefing stage and the subsequent decisions made during the design and assembly phases, affected by the choice of materials and the soundness of the detailing. For many years, running costs were only given superficial attention at the design stage, although this has changed with the use of life cycle costing techniques that help to highlight the link between design decisions and costs in use. Materials and components with long service lives cost more than those not expected to last so long, and designing to reduce both maintenance and running costs is likely to result in an increase in the initial cost. However, over the longer term, say fifteen years, it may cost the building owner less than the solution with lower initial cost. It is a question of balancing alternatives at the design stage and possibly educating the client about building costs in use.

One may divide consideration of costs in use into the basic strategic decisions that need to be taken early in the design, and the more tactical decisions that can be left to the specification stage. The most significant of the former concern the energy consumption of the building. Increased insulation, for example, will reduce running costs at the expense of increased capital cost. The decision on whether or not to install thermostatically controlled radiators, which will involve a similar balance between initial and running costs, may be something that can be deferred to the detailed specification stage.

The cost of demolition and materials recovery is commonly ignored. This is partly because the client may well have sold the building (or died) long before the building is recycled, and partly because such costs are traditionally associated with the initial cost of the future development. Again, this may be of little concern to the current client who is looking for short term gain with minimal outlay. However, if environmental issues are to be taken seriously, then the recycling potential and ease of demolition should be considered during the design phases and costed into the development budget.

Mandatory requirements

There are several factors that can be grouped under the heading mandatory requirements that influence the specification of buildings. The most obvious relate to safety and well being and are expressed in national Building Regulations and laws. The most obvious mandatory requirements relate to structural safety, fire safety, health and safety, and access. Others are linked to specific building types, e.g. banks, schools, hospitals and prisons, and these should have been highlighted and specified in the client briefing documents. Insurance companies also have a role to play here, because the specific requirements of insurance companies have to be taken into account at the design and detailing phase

Performance (technical) characteristics

The manufacturer will describe the product's technical characteristics, and these will need to be reviewed in the context of the overall design.

Quality

Because so much effort is often expended on the design and construction process it is crucial to understand quality from the perspective of the building users. While space and facilities are critical to ensuring user satisfaction, so too are the materials that form the finishes. Quality materials and craftsmanship carry a higher initial cost than cheaper and (possibly) less durable options, yet the overall feel of the building and its long term durability may be considerably improved. Quality controls, the use of quality management and the adoption of a total approach to quality from everyone involved in the construction process will be instrumental in determining the finished quality of the building.

Risk and evidence in disputes

The specification of a building product that may fail in use is perceived as a major cause for concern by design and engineering practices, the majority of which go to great lengths to reduce their exposure by specifying products that are known to them. Mackinder (1980) noted that the partner of an architect's office influenced 'major' decisions regarding product selection, strongest at stages C and D, during which important decisions were taken that influenced the structure and the finished appearance of the building. As the design progressed to the detail design stages, where the number and variety of decisions increased, their importance in terms of both the cost of the job and the overall appearance of the building was seen to be reduced; Mackinder referred to these as 'minor' decisions. With the perceived reduction in the importance of the decision making, it was common practice to involve less senior members of the office, often with varying degrees of supervision from their more experienced colleagues. There would appear to be a hierarchy within the office, with the recently qualified (younger) members of staff, with little experience, supervised by a more senior member of the office, who in turn was supervised by the most senior member of the office, the partner, where major product selections were concerned. A combination of the specifier's experience and their position in the office hierarchy appears to influence the relative importance, as viewed by the partner, of the products being selected.

In interviews with specifiers, their main concern with regard to risk was related to defects in the products they may select. However, in a comprehensive survey of construction problems involving specifications that led to claims, litigation or arbitration deficiency in product performance was some way down the list of problems (Nielson and Nielson, 1981). Top of the list were problems caused by the use of the phrase 'or equal' (25 per cent of cases). This was followed by problems caused by ambiguous writing, differences between specifications and plans, problems regarding buildability, and inaccurate technical data (12 per cent each). Together, these five problem areas accounted for 73 per cent of the cases. Problems with product specifications accounted for only 8 per cent. Caution is required here because America has different specification habits from other countries, but the data are useful in helping to highlight some common problems faced by all specifiers. One may also argue that specifiers who are aware of these figures may exercise more care in ensuring that such problems did not occur.

End of chapter exercises

- Take a small part of your design project and try to determine the most important selection criteria for a particular element; for example, a door or window. Once you have identified the most important criteria try to rank them in order of importance and justify your rationale.
- Select three manufacturers' websites and rate them for
  (a) quality of design
  (b) ease of use
  (c) the level of detail of the information
  (d) the environmental credentials of the manufacturer (objective or subjective?)

  Is the site that provides the best information on the product also the easiest to use? Do you think that the design of the websites might affect your choice of product rather than the suitability of the product itself?
- Could you specify a product (e.g. a door or window) based on the information provided on the website, or would you need to contact the manufacturer for additional information?

5 Writing specifications

Once a decision has been made about the performance parameters required and/or the proprietary products to be used, the specifier has to confirm these intentions in the contract documentation, i.e. they have to be 'specified'. This can be done by referring to the product on the drawings, by including it in schedules of work, or through the written specification. The written specification is an important document in ensuring a quality building for which adequate resources must be allocated by the office manager (see Chapter 6). The specification must be well written, comprehensive and free of errors, a task that requires considerable time. It is an essential part of the design process that requires particular skills in researching different product characteristics and being precise in communicating those requirements to a variety of organizations and individuals.

Writing the specification is an evolutionary process, initially in an outline form based on the briefing documents, with the level of detail increasing as the design develops. The final act is the completion of the written specification. On small and relatively simple projects it is common for the architect or engineer to develop the specification as the design develops, with little or no assistance from others. Thus, the continuity of thought is maintained from early design ideals, through increasing levels of detail to the confirmation of requirements in text and drawings. Given the designer's familiarity with the design it may be possible to address the task of writing the specification as a separate activity from the design work.

For larger and more complex projects it is common for the design architects and engineers to hand over the tasks of detailing and specification writing to others within the office. Thus, the specification writer may have a small role in the product selection process, writing the specification based on information provided by the design team. In this case it is critical that the interface between the conceptual designer(s) and the specification writer(s) is as seamless as possible. This calls for excellent communication between individuals and appropriate managerial procedures to ensure that the design intent is transferred to the contract documents. This often involves a stepped process, moving from a preliminary project description to the outline specification and then on to the final written specification. This stepped approach helps to stop essential decisions and information from being forgotten and allows the design manager and client to review and sign off the specification at predetermined points in the design programme, in accordance with quality assurance procedures.

In the preliminary project description it is usual to list major building materials, assemblies and components that complement the initial design ideas and values. These may be listed as a set of preliminary performance requirements and/or a preliminary list of manufacturers/suppliers to be investigated further as the design develops. These are sometimes set out under major headings, such as superstructure, interiors and services and sometimes under the more detailed Common Arrangement of Work Sections (CAWS) headings, from A to Z. These preliminary lists and descriptions should reflect constraints of budget and time as set out in the briefing documents. In situations where the client possesses a client specification, this will form a significant part of the preliminary project description.

As the schematic design develops and the amount of detail starts to increase, the preliminary project description will be tested and revised. During this period (which may be lengthy on large projects) the design team will convert the information in the preliminary project description into an outline specification. The outline specification will follow the same format as that to be used for the written specification that will form part of the contract documents. Typically, this will be arranged under the CAWS headings. The compilation of the outline specification is an evolving process, with various elements developing at different stages to reflect the iterative nature of the design process and the degree of completeness of plans, sections and details. Thus, the outline specification will contain a wide range of information, some of which will be almost complete and some will be relatively vague because the appropriate level of detail in the design has yet to be reached and agreed. The outline specification will list items such as materials, products, finishes and performance standards, as well as specific fabrication and installation criteria. By preparing such a document it is possible to identify any unusual requirements at an early stage, thus allowing the design team to check that they are achievable within budget, time and quality parameters before confirming them in the written specifications. The ability to identify unusual and sometimes unachievable requirements (e.g. relating to delivery and installation) will result in a search for alternatives and possible redesign of particular elements. Obviously, the earlier this is done the less rework will be required later in the process. The outline specification forms a useful checklist for the design team, and assuming its currency is maintained the document can aid communication between participants working concurrently on the design. This document can also be very useful as a basis for revising and refining cost estimates as the design develops, providing the basis for informed decision making.

The outline specification will normally be evaluated during the formal design reviews and value/risk management exercises (if used). Areas of uncertainty will need to be resolved in the detailing phases and then confirmed in the written specification. As the outline specification is developed the design team will collect a considerable amount of technical information relating to specific materials and products, and this file of information will provide an excellent resource for the specification writer. This stepped approach helps to minimize the time required to search for technical information at the specification writing stage, given that much of the searching and decision making should be complete.

The written specification will be developed and read in conjunction with drawings and schedules and used by people from different backgrounds for a number of quite different tasks. The requirements and standards determined during the briefing process will be used to develop both the design and the detailed specification during the pre contract period. This information will then be transmitted to site, and will form a record of the work for future reference. Good specification writers are aware of the different uses to which the documentation will be put and the different backgrounds and requirements of those likely to read it. More specifically:

During the pre contract phase, the document will be:

- developed from, and be central to, the briefing process
- used by the design team as part of the concurrent design development (and elements of the specification may be submitted as part of the town planning and building control approval stages)
- read by the quantity surveyor/cost consultant to prepare cost estimates and the bills of quantities
- read by the contractor's estimators to prepare the (tender) price.

The written specification must be complete at the end of this phase. Discrepancies identified by the cost consultant and the contractor's estimators should have been resolved, since changes made after the contract has been awarded will have cost and resource implications.

During the contract, the document will be:

- read by the contractor's agent on site
- read by sub contractors and site operatives
- read by the project manager/contract administrator, clerk of works and/or resident architect/engineer to check that the work is proceeding in accordance with the contract documentation.

Changes made during the course of the contract need to be recorded and immediately following the completion of the contract the written specification (and associated contract information) should be revised. This ensures that the building owners are provided with accurate 'as built' documentation.

Post contract, the document will be:

- used as design record of materials used and set standards
- possibly used as a source of evidence in disputes
- used as a source of information for maintenance, facilities management and recovery management
- analysed by the design team for feedback of knowledge into the master specification and office procedures.

Writing the specification

The primary aim of the written specification is to convey information to the reader that cannot easily be indicated on the drawings and schedules. The contents of the document will be concerned with the quality of the materials and the quality of the workmanship, neither of which can be shown adequately on drawings. Although a large proportion of the document will be common to many different projects, specific details will be dependent on the nature of the particular building being specified and the design philosophy of the designer or design office. As noted above, writing the specification involves considerable knowledge of construction, materials and working methods, in addition to good writing skills, if this practical document is to be of value to the reader. A guide to specification published in 1930 noted that a complete knowledge and understanding of the details of building construction, properties and cost of materials were paramount. Until these were mastered, it would not be possible to write a specification (Macey, 1930), a sentiment that still holds true today.

Badly written specifications will result in claims for extras from the contractor. It is also highly likely that the site personnel will not bother to read the document if they find it to be deficient in any regard. Willis and Willis (1991) claim that there are two essentials to specification writing: to know what is required and to be able to express such requirements clearly (echoing Macey's earlier advice). They note that many specifications fail because of shortcomings in the first stage. There may have been insufficient thought and/or insufficient knowledge of building construction.

- *Insufficient thought*: to accuse professionals of insufficient thought is a serious matter. Professionals have a duty of care to their clients and are expected to behave in a 'professional' manner. They should be capable of applying sufficient thought and sufficient knowledge to the problem in hand. However, insufficient thought may be given to a specification more because of a lack of adequate resources, especially time, rather than because of simple incompetence. It matters little how knowledgeable or how good an individual is if inadequate time is allowed to consider the options carefully, to make an informed decision and confirm this decision in writing within the specification. Moreover, sufficient time must also be given to considering the clarity and accuracy of the wording used.

- *Insufficient knowledge of building construction*: at first sight, one may assume that insufficient knowledge is only a problem for those just starting work in the building industry. It was noted earlier in the book just how little time is spent teaching the art of specification, and many observers of architectural education on both sides of the Atlantic have become increasingly critical of the small amount of technical instruction within architectural programmes. Because of this, young practitioners have to learn through experience in the design office and thus require close supervision and support in their early years. However, problems of insufficient knowledge can also arise when an experienced designer is faced with a new type of building, a new problem or new products. Again, time is required for the individual to acquaint himself or herself with the new information before informed decisions can be made and the specification accurately written.

Both of the essentials identified by Willis and Willis can be addressed through:

- adequate time to complete the task, i.e. good programming and management
- easy access to current and relevant information
- employment of experienced staff
- close supervision of less experienced staff
- continual education and training.

Naturally, to express requirements clearly in the specification, the writer has to know first what is required and secondly what goes where. Loose thinking will inevitably result in loose writing. Vague specifications are often an indication that the decision making process had not been resolved at the time of writing the specification. Bowyer (1985: 11) sums it up rather neatly:

> Unless the designer knows what he wants he cannot expect either the specification writer to describe it, the estimator to price it or the builder to construct it.

In well managed organizations, the writing of the specification and any alterations to it will be covered by quality assurance and quality control procedures that should be designed to eliminate errors and omissions. However, the management of many design organizations fails to live up to this ideal, with quality standards determined by the whim of the senior partner(s) rather than by any written documentation. What is too frequently not realized is that quality-control procedures are an important factor in attaining and maintaining a competitive advantage. Information for building design is produced and used by organizations that are in business to make a profit. Regardless of their size or market orientation, organizations must give their clients, i.e. their customers, confidence in the service that they provide. Those involved in the design and construction of buildings must also satisfy their clients with the quality of the finished building. A consistent approach is required, and quality management systems may be seen as an essential tool in ensuring consistency.

Specification writers

The production of written specifications has undergone major changes in the past few decades, with computer software and standard formats making the editing, reproduction and transmission of the document much easier than it used to be. However, we should remember that specification writing is a skill, and the success of the written specification will depend very much on the abilities of those involved in its production and the constraints under which they work. It may seem like a statement of the obvious to say that to be usable, written specifications must be easy to read, but they do not always have this virtue. In the USA, there is a separate profession of specification writers, i.e. it is carried out by qualified professionals who dedicate all of their time to the task. One can see from this that the task is seen to have greater importance, and the role carries greater prestige than it currently does in the UK. This probably reflects the greater tendency for post contract litigation in the USA, where the professional specification writer is seen as a defence against legal action. Some large offices in Britain employ people to write and check specifications, leaving the designers to design and the managers to manage. Such an arrangement calls for close coordination between designer and specification writer. However, the majority of British design practices are small, and so the specification is often written by the same person who carried out the design and detailing of the scheme, which means that they only spend a small amount of their time dealing with this task. In smaller offices, the designer has little option but to write his or her own specification. The point being made here is that the majority of specification writers are 'part time', and therefore it is particularly difficult for them to be expert at this task, but expert they must be because there will be legal implications if the specification is wrong.

From the authors' experience of design management, it is clear that some people are well suited to specification writing, whereas others are not. Good specification writers tend to be individuals with a very good technical ability and considerable experience of both design and building operations. They should be precise and have an eye for detail. They also have to be exceptionally good at interpreting designers' drawings, i.e. they need to be aware of the project's design goals. It goes without saying that they must also be able to communicate in writing.

Because of their training and background, many designers prefer to use graphics to communicate in preference to writing and, as a result, may not be particularly adept at writing good specifications. Thus, those working in very small offices who have little option but to undertake this activity need to learn this skill. Ideally, specification writers need to have excellent knowledge of construction materials and products, construction methods and project management. They should also be aware that disputes arising from errors in the contract documentation will have serious legal consequences, so attention to detail is paramount. Once the requirements have been established, the next step is to express them clearly, concisely and in a logical manner in a written form.

Typical job advertisements for specification writers in the professional press help to highlight some of the required characteristics. These advertisements usually state that the role is an 'important' or a 'key' position within the design office and imply, or request, a certain level of experience and maturity. Summarizing the contents of a few advertisements, it is evident that applicants would typically need to have the following:

- (extensive) experience of writing specifications, including familiarity with commonly used specification formats (e.g. NBS) in addition to office standard specifications
- the ability to interpret and implement regulatory and technical data
- knowledge of national, European and international standards, codes of practice and legislation relating to construction
- working knowledge of a range of construction contracts
- extensive knowledge of materials, construction methods and systems used in a range of buildings (both new build and refurbishment)
- the ability to communicate effectively with a wide range of project participants
- familiarity with (named) information technology systems and software.

Adopting a systematic approach

The order in which the specification is written will be determined by the characteristics of the individual and any office procedures. The writing and editing sequence may also be influenced by the amount of information available at a particular time (i.e. what can be specified and what cannot), together with any items that need to be decided and confirmed before other decisions can be made. Critical is the ability to approach the task in a logical and ordered manner, completing one section before moving on to the next. This is likely to involve some interaction between the specification writer and the design team to resolve discrepancies and omissions in the drawings. It is good practice to specify the work required before attempting to complete the materials and workmanship clauses, i.e. to identify what needs to be done and then what materials and methods should be used to achieve the task. Preliminaries are often the last section completed because they are largely independent of the work sections. Consistency is the key, and this is helped by using industry-standard formats or office masters. In situations where a non standard format is used, it is essential that the document is set out in a logical manner, complete with an index of main headings to guide the reader. When the specifier has finished the writing and editing tasks the final step is to check the document for completeness before it is released for use by others.

Guides and guidelines

A small number of guidelines and checklists is available. The aim of the guidelines is to help the specifier to edit the various sections in the written specification. The guides are particularly useful for those new to the act of specification writing and those using an unfamiliar format for the first time. The commercial suppliers of national specifications also provide guidance documents and design offices usually provide some form of guidance for editing the office master specification. The commercially available standard specifications available on line have guidance notes built into the document. Checklists are used to prompt the specifier in making a decision. These vary from a list of basic information that is needed to prepare the document through to very detailed lists of dos and don'ts.

Standard formats

Well managed design offices recognize the importance of consistency, and the majority have developed standard formats for a variety of tasks, including the written specification. Standard formats can save an enormous amount of time, but care should be taken to ensure that errors are not being transferred from one project to the next through lazy copying without due care (see Chapter 6). Over time, different countries have developed their own 'standard' specifications as a means of guiding specifiers and also in an attempt to bring some form of consistency and common documentation to the building process, examples being America's Construction Specifications Institute (CSI) SPECTEXT, Australia's NATSPEC, the Netherlands' STABU and the UK's National Building Specification (NBS). Although these are widely promoted and used, other specification packages are available from commercial suppliers, in addition to which large design organizations tend to use their own particular systems. As a result, these 'national' systems have never become universally used.

There is a powerful argument for everyone in construction to use the same system, although this ideal does not seem attainable in the foreseeable future.

National systems are a useful tool in ensuring clear, concise and (more importantly) familiar documentation for individual projects. In addition to their economic advantages to both designer and contractor (and hence client), there should be less likelihood of ambiguity and confusion caused by unfamiliar-looking documentation. Assuming, for one moment, that everyone in construction could agree to a national format for specifications, this would form part of a quality assurance package for the industry. One might be tempted to suggest that some of construction's persistent ills could be eased by such an approach. Perhaps this is wishful thinking because there are also arguments against this approach. National standard specifications do tend to be rather complex, and the format and defaults may not be to everyone's taste. Time and skill are required to understand them and apply them correctly, so that they are just as open to misuse and abuse as other systems. The two most common problems relate to the inappropriate editing of standard clauses by inexperienced, overworked or lazy specifiers. Either almost every clause is included 'just in case', or conversely the document is severely edited, resulting in the omission of important information. In both cases the resulting written specification is unlikely to be of much use to anyone. Another problem associated with national systems is the design manager making the assumption that several specifiers can contribute to the specification as they develop the design. This can sometimes work, but in the majority of cases the resulting document is inconsistently edited and hence confusing.

National Building Specification

The NBS is widely used in the UK. This suite of specification formats includes *NBS Building*, *NBS Engineering Services* and *NBS Landscape*. This commercial specification package helps to make the writing of specifications relatively straightforward because prompts are given to assist the writer's memory. Despite the name, the NBS is not a national specification in the sense that it must be used: many design offices use their own particular hybrid specifications that suit them and their type of work. Moreover, the NBS is only available via subscription to the service providers, but this ensures that the user has access to a document that is kept up to date through regular revisions.

NBS Building is available in three different formats to suit the size of the particular project, ranging from Minor Works (small projects), to Intermediate, and Standard (large projects). It is an extensive document containing a library of clauses. These clauses are selected and/or deleted by the specifier and information is added at the appropriate prompt to suit a particular project. The NBS ensures a consistent format, and the provision of prompts reduces the danger that some element is forgotten. However, as with all templates, the quality of the finished specification still depends on the ability of the individual to fill in the gaps, delete the unnecessary clauses, retain the clauses applicable to the design and add the necessary detail.

Office master specifications

Office master specifications differ from the commercially available standard specifications in that they are designed to suit the unique characteristics of a particular architectural or engineering office. These bespoke documents are often developed from a standard format, such as the NBS, and adjusted to suit the unique characteristics of the office's client and project portfolio. The major benefit is that the master specification can be revised to incorporate knowledge harvested from individual projects. It is also much quicker for specification writers to apply the office master specification to new projects than it is to start from scratch with a commercially available template. These savings in time need to be offset against the challenge of keeping the master specification document up to date to reflect developments in standards, products and practices. Failure to do this on a regular basis will result in a poor quality document. Maintenance and updating can consume considerable staff time, and hence office master specifications are best suited to large and highly specialized offices. For small offices and offices that have a diverse client base and engage in a wide variety of work it is usually a better option to subscribe to a commercial supplier of standard specifications. These and associated issues are addressed in Chapter 6.

Shortform specifications

Standard specifications tend to be rather long documents and for some contexts it may be preferable to use an abbreviated format, known as a 'shortform' specification. Shortform specifications, as the name implies, are designed to be much shorter documents than a standard specification, and hence quicker to read and easier to use. The term 'sheet specification' is sometimes used, since the specification is written on a few sheets of paper. Written well, these documents will have been reduced to their shortest length, without compromising their effectiveness. The approach is to use a limited set of article headings and to keep the level of detail as broad and as simple as possible, thus reducing the amount of text within the document. This can be achieved by using the reference specifying method and proprietary specifying, with reference to a list of schedules, tables and drawings. The descriptive method and the performance method are less suitable since they both require extensive use of text. It is also possible to adopt a modified (shorthand) style of writing to reduce further the amount of text required. However, specification writers should make sure that their intended audience is also familiar with the shorthand language and terminology if problems with misunderstanding are to be avoided.

Although the concise text may suit the users of the written specification, it is generally acknowledged that editing the specification and revising the language to suit the short format is very time consuming and can take as long to produce as the more usual format. As such, shortform specifications tend to suit small, relatively simple projects of limited scope and a high degree of repetition, for example the refurbishment of a large number of residential units, commercial internal shop fitting and simple light industrial building shells. Shortform specifications are not suitable for contractual arrangements that rely on competitive tendering, since the broad level of detail makes it impossible to compare tender submissions objectively. However, they do suit relational forms of contact, such as partnering, in which the cost of the work is negotiated with the contractor. They also suit design and build type contracts, where the contractor requires a specification that is not too specific and hence unnecessarily restrictive.

Shortform specifications demand a considerable amount of specifying experience and writing skill and they should not be tackled by inexperienced specification writers. Standard formats are available from commercial suppliers. The general principles described below still apply.

Specification language

Specifications have their own language a language that takes the inexperienced specifier and contractor some time to understand and become familiar with. However, there are guides, and standard formats can go a long way towards achieving consistency. For example, the CSI's *Manual of Practice* (CSI, 1996) provides comprehensive guidance on specification language. Consistency of word usage is the key to a well written document and, if confusion of the reader is to be avoided, the writer must take care always to use the same word or phase in the same sense throughout the document. For example, one of the most common causes of confusion can arise when specifying sizes of components; all sizes should be either 'unfinished' or 'finished' sizes. Where it is necessary to use both conventions, it is important that the specifier makes it clear to the reader whether each size specified is 'finished' or otherwise.

Standards and Codes of Practice can, if used carefully, help to shorten the description of a particular material. The extent to which British (BS), European (EN) and International Standards (ISO) and Codes of Practice are included in a specification is often a point for debate. As discussed in Chapter 3, no one designer will have a working knowledge of all the relevant standards and codes; nor, for that matter, will the contractor. It is the specialist sub contractors who will know and understand the standards and codes that apply to their particular area of expertise. Many of the standards have sub divisions relevant to the same material, and so the specifier must make sure that the reference number quoted includes the correct sub division. A good rule of thumb is to specify only standards with which the specifier and/or the design office are familiar. Unfortunately, it is common practice to quote general standards without taking sufficient care (because of a lack of time or simply through laziness) to ensure that it applies to the particular job, and the specification then becomes somewhat meaningless. This is a particular problem with rolling specifications from one job to the next and, to a lesser extent, with the office master specification, if not maintained on a regular basis (see Chapter 6).

Imperative or indicative mood

The manner in which a specification is written is important and can help in reducing repetitious and tedious sentence structures. Specifications can be written in the imperative mood or the indicative mood.

- *Imperative mood*: the imperative sentence is concise and easily understood because the verb that defines the action forms the first word in the sentence. The reader is directed by verbs such as Apply, Install or Remove. The imperative mood is seen as the major factor in producing clear and concise text; it makes it quicker to write (and read) and hence saves time and money. For example: 'Apply two coats of emulsion paint to ...'.
- *Indicative mood*: the indicative mood is written in the passive voice. Sentences require the monotonous use of 'shall' and can be unnecessarily wordy. For example: 'Two coats of emulsion paint shall be applied to ...'.

Of the two the imperative is preferred for its clarity by both the authors and others (e.g. Cox, 1994; CSI, 1996; Rosen and Regener, 2005).

Style

Proper style will ensure clarity, brevity and accuracy. The style of writing will be influenced by the level of detail required to describe adequately the required attributes, which should be appropriate to the type and complexity of the project. Although written specifications form part of the contractual documents and hence are legally enforceable, writers should avoid the use of 'legalese' and unusual terminology. Similarly, writers should avoid the urge to engage in creative and over elaborate prose. The style adopted should be consistent within the design office regardless of the degree of detail required for a particular project. Specification writers must have the ability to express their intentions in plain and grammatically correct English. Writers should use:

- short sentences
- simple sentence structure
- plain words and terms.

Abbreviated words, acronyms, symbols and slang terms must be avoided since they may be misinterpreted. Care must also be taken with some technical terms, especially if the specification is to be used by people from a different region or culture from that of the specification writer. For example, some of the building terminology used in Scotland differs from that used in England. Spelling must be correct and consistent throughout the document (e.g. use 'window cill' or 'window sill', not both). Punctuation must also be used correctly. If in doubt, use two short sentences, not one long one. Take particular care with numbers and make sure that the conventions used in the written specification are the same as those used on the drawings.

Certain terms and phrases should not be used in written specifications (or on drawings). Some of the more obvious terms to avoid include 'as appropriate' (to whom?), 'as indicated' (where?), 'as intended' (here we are into the realms of guesswork) and 'as approved' (by whom, when and how?). One of the authors' favourite phases is 'to the satisfaction of the architect (or engineer)', which usually amuses contractors, but makes it impossible to price the work since the phrase is both subjective and ambiguous. Another favourite, and very common, phrase for work to existing buildings is '... to match existing'. For the majority of existing buildings, a directive such as 'brickwork to match existing' may be impossible to comply with, simply because the exact brick is no longer produced, and so a brick that is a very close match is used. Moreover, such descriptions invariably forget to mention items that may not be apparent to the contractor. The specifier may be aware that lime mortar is used in the original work, but this need not be apparent to the contractor, who may assume the use of a cement mortar.

Contractors are also quick to complain about written specifications that contain 'catch all' clauses. An example would be to ask the contractor to 'supply everything necessary for the satisfactory completion of the work'. Such instructions are meaningless to those trying to carry out the work and are an indicator of inexperienced (or lazy) specification writing. Similarly, a specification peppered with references to too many standards may be difficult to deal with on the building site.

Clarity, brevity and accuracy

The specification must be written in such a manner that it conveys the intentions of the designer to the contractor. This may appear to be an obvious statement, but specification writers must constantly bear in mind the fact that readers of the specification will not have been party to the decision making process that led to the contract documentation. The readers can only read the documentation to see what is required of them. As mentioned earlier, there is a need to be able to express one's intentions clearly, for which sufficient thought and sufficient knowledge of construction are prerequisites. In many respects, the use of 'standard' specifications has helped individuals to write specifications because the format is already supplied. The writer then has the relatively simple task of deleting the clauses that do not apply and adding information as appropriate.

Designers have their own way of working, and many have 'golden rules' that they apply when designing, detailing and writing specifications. In offices where managerial control is not particularly good, this can and does lead to information taking a variety of slightly different forms, reflecting the idiosyncrasies of their authors. The end result can look unprofessional, can lead to confusion and, in the worst case, can result in errors on site. Professionally managed design offices take a much more considered and controlled approach. Designers work to office standards of graphic representation and to a standard approach to detailing, product selection and specification writing. Guidance for members of the design organization is provided in the office quality manual. Six golden rules may be suggested:

1. *Clarity and brevity*: the most effective information has clarity and is concise. This is far easier to state than to achieve because it is impossible to represent everything in an individual's mind on a drawing or in text. The skill is to convey only that which has relevance and hence value to the intended receiver. This can be a matter of knowing when to stop writing. This will help the receiver to avoid information overload and enable him or her to concentrate on the relevant information without unnecessary distraction.
2. *Accuracy*: it is important to be accurate in describing requirements because confusion will lead to delay and errors on site. Use correct words to convey exact instructions, use correct grammar, units and symbols, and avoid ambiguity. Words and symbols should be used for a precise meaning and be used consistently for that meaning throughout the document. Instructions should be given accurately and precisely. Use a limited vocabulary of words. The document should be complete: do not leave out important information or leave clauses partially completed.
3. *Consistency*: whatever the approach adopted by the design office and the individuals within it, it is important to be consistent, both in the meaning of words and in the approach to specification decisions. For example, if specifiers do have individual and different attitudes to detailing, those on the receiving end should be able to interpret instructions as long as the approach remains the same. Use of graphics, dimensions and annotation should be reassuringly consistent across the whole of the contract documentation. Both

computer aided design (CAD) packages and the use of the CI/SfB should help to achieve this goal.

4. *Avoiding repetition*: describe items once and in the correct place. Repetition of information in different documents is unnecessary, is wasteful of resources and, when repeated slightly differently (which it invariably is), can lead to confusion. Repetition, whether by error or through an intention to help the reader, must be avoided both within and between different media. Eliminate unnecessary words and sentences in the written specification and avoid notes on drawings wherever possible. Be concise.
5. *Redundancy*: there is always a danger that superfluous or redundant material will be included in a project specification. Text from the master specification may be redundant because it is not relevant to a particular project. Rolling specifications from one project to the next invariably results in redundant text. A favourite example of the authors comes from a large refurbishment project where the specification said 'Remove defective render...'. There was no render on the project. In this example, the specification had been rolled from another project that did have rendered walls. The document not only becomes larger than it should be, but will lead to confusion and may well undermine the credibility of the written specification (and those who contributed to it). Careful editing should help to remove the majority of redundant material.
6. *Checking*: check and double check for compliance with current codes and standards, manufacturers' recommendations, other consultants' details and compatibility with the overall design philosophy. Common problems encountered by site personnel can be reduced significantly through a thorough check before information is issued to the contractor. Not too long ago, it was common for drawing offices to employ someone to check all drawings and specifications before they were released from the office. Unfortunately, in the constant drive for efficiency and ever tighter deadlines for the production of information, such checks have been left to the individuals producing the information. Self checking is suspect and subject to error simply because of the originator's over familiarity with the material. Managerial control is essential in this regard and must be costed into fee agreements. Checking for omissions and errors before the written specification is released from the specifier's office will, over the course of the project, be time well invested.

Typical contents of a written specification

All specifications need to be arranged in a logical order to allow the specification writer and the users to navigate their way around the document. A common feature of national specifications is to divide the work into sections and then to subdivide these sections further to accommodate increasing levels of detail.

Bespoke specification documents will reflect the idiosyncrasies of the design office, specifiers and projects. So, although specification formats differ (e.g. in the way sub divisions are ordered and numbered), the main layout of the document should be relatively familiar to the users of the document. In the UK it is common practice to arrange the contents of specification under the CAWS headings from A to Z, as shown in Table 5.1.

Specifications for building, whether for new or alteration works, have two main types of clauses. First are those that describe the general conditions under which the work should be carried out and various obligations of employer, contractor and designer. These are grouped under the CAWS headings of A–C

Table 5.1 Common Arrangement of Work Sections (CAWS)

A	Preliminaries/general conditions
B	Complete buildings/structures/units
C	Demolition/alteration/renovation
D	Groundwork
E	In situ concrete/large precast concrete
F	Masonry
G	Structural/carcassing metal/timber
H	Cladding/covering
J	Waterproofing
K	Linings/sheathing/dry partitioning
L	Windows/doors/stairs
M	Surface finishes
N	Furniture/equipment
P	Building fabric sundries
Q	Paving/planting/fencing/site furniture
R	Disposal systems
S	Piped supply systems
T	Mechanical heating/cooling/refrigeration systems
U	Ventilation/air conditioning systems
V	Electrical supply/power/lighting systems
W	Communications/security/control systems
X	Transport systems
Y	Services reference specification
Z	Building fabric reference specification

(Table 5.1). Second are those that describe the materials and workmanship required in detail. This section forms the main body of the specification and for new building work would be set out under CAWS headings D–Z (Table 5.1). The work sections are further sub divided to create a three level system.

The specification clauses applying to each of these detailed aspects of the building will follow from the details of the design and the materials and components selected, but there are some aspects of the specification document and some classes of work that require special comment. Note that contracts involving alterations and conservation works are more complex and need to be approached differently.

Materials and workmanship clauses

The clauses in a specification are there to limit the choices open to the contractor and to ensure specific requirements are met, and these will be decided by the designer to meet the requirements of the design. In the final analysis the quality of the final product depends on the quality of the design, the components and materials used and the workmanship. The last two of these present something of a challenge to the specifier because any definition of quality should implicitly include the measure by which it is to be judged. Phrases such as 'the very best quality' might have been of some use in past centuries when everyone agreed upon what they meant, but are of no use today. If necessary, the grade of concrete aggregates may be tested with sieve tests and it is possible to test paints to ensure that they have not been unduly thinned. If the specifier has any doubts about the quality of material that might be used then he or she should consider what tests might be applied in the case of doubt or dispute, and include them within the specification. Care should be taken to ensure that any such clauses are both relevant and precisely worded.

If tests are to be specified it is essential to ensure that there is a mechanism for carrying them out, reporting the results and taking action should the test fail. A slump test on ready mix concrete is simple to carry out and if the batch fails the test it can be immediately rejected. A failed cube test on concrete raises more difficult questions because the concrete will have been in place for some time before the results are obtained. It is also important to ensure that there is someone competent to carry out the tests if they are to be meaningful. One of the authors was supervising work where the contractor was to carry out slump tests on the concrete as it was delivered. On the day that the first batch of concrete was delivered he appeared with a brand new cone and required instructions and a demonstration of how to use it.

There is no substitute for a properly trained, conscientious workforce appropriately managed to obtain good workmanship, but these attributes are largely outside the control of the specifier. Ensuring standards on site is a matter for supervision and inspection. Where the quality of surface finishes depends on standards of workmanship it may be appropriate to require test specimens to be made for approval. The appearance of the test specimen then becomes the standard with which the final product needs to comply and which will be referred to in the specification. This approach can be adopted, for example, with concrete work that has a bush hammered or exposed aggregate finish. Clearly, a similar approach can be used for specialist interior decorative work. This method is also used for conservation work (see below).

Specifications for alteration work

When dealing with alterations to existing buildings, it is sometimes easier to subdivide the specification according to certain areas of the building and/or according to particular rooms. By composing the specification in this manner, everything connected with a particular area is described together and is generally preferred by the personnel on site. On larger alteration projects, it may be easier and more efficient to use a mixture of both methods, with new work classified by work section and alteration work by room. The layout of the specification for alteration work will depend on the size and complexity of the job and the format adopted by individual design offices to suit individual projects.

Specifications for conservation work

Conservation work raises a number of particular problems: the extent of the work and the methods to be used are commonly constrained by the need to retain historic fabric, both the skills and the materials used may be obsolescent and therefore only obtained using specialist contractors, and the designer and specifier may have to meet the requirements of the conservation officer with whom the details of the specification will have to be agreed. Each of these issues needs to be considered.

The extent of the work and the methods to be used may have to be agreed with the local authority's conservation officer. In England this may also involve advice from English Heritage, while Scotland and Wales have their own advisory bodies. In the USA one may have to deal with controls varying from city ordinances to Federal regulations depending on both the historic and commercial status of the building. In other words, there may be specific local controls that govern historic buildings that the designer and specifier will have to consider, and there may well need to be some negotiation with these authorities before a final specification can be drawn up. The authors take the view that this is best done by architects or other professionals with experience of this kind of work and a sufficient understanding of conservation issues to be able to negotiate with the conservation officer, as more than one approach may be possible.

It must not be forgotten that this is the client's building. With new building work the designer will have determined the client's requirements before starting the design. With conservation work the client's intentions and the overall strategy of repairs need to be agreed with the client. In many cases clients may be unfamiliar with conservation issues and so this often means explaining the implications of optional strategies to them.

Both the materials and the trade skills needed for conservation work are those of the past and not those commonly available today, and this presents problems of specification and quality control as well as availability. Materials used in the original building may no longer be available or only obtainable from specialist suppliers. There is also the difficulty of specifying mixes used in the past, finding the skills to work with such materials and controlling the quality of such work. It is still possible to mix the daub used in wattle and daub, although very few contractors will have had experience of this. In such cases, the specifier may need to include descriptions of operations that would otherwise be unnecessary. For example, when brickwork is laid with cement mortar, the joints are struck as the bricks are laid. If lime mortar is used, this must be carried out as a separate operation.

Rather than referring to modern standards and specifications, the specifier may wish to refer to those in earlier trade manuals. A useful source here is the annual *Specification*, which provides specification clauses used in work from the end of the nineteenth century. However, this does not solve the problem for earlier methods of construction or the problem of quality control. Lime ash was once used both for rendering and for laying floors but, while the materials are available today, the mixing and applying of this material require an uncommon skill and few will have had the necessary experience. The more basic sand lime renders and mortars are more familiar to present day contractors, but there is the temptation for them to adulterate these with cement so that they harden more quickly. In one case where a render had failed, the surveyor had specified a very weak lime mix. He claimed that this had always proved satisfactory in his previous contracts, but it was more likely that contractors had been adding cement to the mix without his knowledge.

The problem of clauses such as 'to match the original' has already been discussed briefly, and this is where the use of samples to be made up by the contractor for the approval of the specifier, and sometimes the conservation officer, is advisable. Such samples are often necessary because an inexperienced contractor may simply not be aware of what is required. There was a case where the architect had to ask the contractor why he thought that the strap pointing that he had used on the sample matched the original struck pointing.

It is worth acknowledging that some specialist contractors will have more knowledge of these earlier building methods than the professionals who have been engaged to design and manage the work. In such circumstances it is worth drawing on their skills. This can be done in two ways. For one contract one of the authors relied on the judgement of the roofing contractor to determine whether or not individual rafters were adequate. This was a situation where exceptionally heavy roofing slates were used, but where the conservation officer wished to retain as many original rafters as possible (Yeomans and Smith, 2000). In specifying repairs for timber structures the authors have also adopted the practice of having a specialist carpenter appointed as a consultant to advise on the types of repair to be used. This is because the carpenter will have a better knowledge of the practicalities of different repair strategies.

This is a specialist area that cannot be dealt with in detail in this book, but one aspect of conservation work stands out and that is the possible tendency to over specify. This comes from either the imposition of unnecessarily stringent conservation requirements or unnecessarily complex repair methods. The former sometimes comes from the requirements of conservation officers because, sadly, not all are familiar with building methods and have been known to specify work that is all but impossible to achieve. This needs to be resolved by negotiation with the conservation officer. The latter may be because designers have insufficient knowledge of repair and restoration techniques. The solution to this is a little humility, being prepared, where appropriate, to take advice from specialist contractors on the methods to be used.

It is clearly important in this kind of work to carry out as thorough an initial survey as possible if the specification for the work is to be adequate. Nevertheless, no matter how thoroughly this is done there is always the possibility of discoveries during the actual execution of the work that call for changes in the design and/or the detailed specification. The professionals involved need to be aware of this and to be prepared to respond when such discoveries are made, and the client needs to be aware at the outset of this possibility and to be informed when any changes are made.

Green specifications

The National Green Specification (NGS) is an independent organization, partnered by the Building Research Establishment (BRE), to host an Internet based resource for specifiers. It provides building product information plus work sections and clauses written in a format suitable for importing into the NBS; thus, in theory at least, making it easier for the busy specifier to select green products. However, Spiegel and Meadows (2006) note that it is a 'formidable task' to ensure that the (green) specification is implemented as intended. In addition to warning specifiers of the pressures to change materials and products during the tendering and building processes, they provide some sample specification clauses for the written specification, to try to limit the extent of unwanted changes (product substitutions). This is also addressed in the NGS clauses, which attempt to make it more difficult for the contractor to propose changes to the specified products. With contractual forms such as design and build and management contracting it is not uncommon for cheaper or more readily available materials and components to be substituted for those specified in the written specification. This has implications for the quality of the completed building and its environmental impact. The pressures to change the specified products may be more acute with green products given their unfamiliarity to contractors, but as we shall see later, this tends to apply to all products.

Prime costs and provisional sums

In situations where it has not been possible to define everything to be specified, the designer can include prime costs and provisional sums. Prime cost (PC) sums will be included in the tender for goods to be obtained from a nominated supplier. This sum will be adjusted against the actual cost of the products selected. For example, bathroom suites or similar items are often included in the specification as a prime cost simply because the client has not yet decided what style and colour are required. Provisional sums are used to cover work and/or items for which insufficient information is available at the tender stage and which cannot be measured or priced accurately, as such provisional sums are particularly useful for alteration works. Another use for provisional sums is to cover the work to be carried out by statutory authorities and utilities companies, such as the connection of mains drainage, water, gas and electricity. Prime cost should also instruct the contractor to include a sum for his profit and attendance. Provisional sums are 'whole element' sums and profit and attendance should already have been included in this sum by the specification writer, so no addition to this sum is necessary by the contractor.

Contingency sum

Building projects will, to lesser or greater extents, have a number of unknown elements that only become known as the work on site progresses. Typical examples are unexpected difficulties with ground conditions and unexpected problems when existing buildings are opened up. The contingency sum is essentially an undefined provisional sum of money to be used if required. Contingency sums can only be used as instructed by the contract administrator. Any sums not used will be deducted in whole or part from the contract sum. Contingency sums are provisional sums and do not have any additions by the contractor.

End of chapter exercises

- Take a particular element of your design project, e.g. a door or a window, and write a specification for it using the following methods:
  (a) proprietary specification
  (b) performance specification.
  Which was the easier to write and why?
- What would you do if, when writing a project specification, you found discrepancies between the information contained within the detailed drawings and the office master specification?
- What are the implications of issuing a written specification to a contractor that is not complete?
- Your office has decided to pursue a more environmentally aware approach to the specification of buildings. What are the implications for writing project specifications?

6 Managing the specification process

Specifying is essentially a task undertaken by individuals within an organizational framework, and however small or large the project there is a need to manage the specification process. This aspect of design management is concerned with the general management of the design office and the management of individual projects within the system. The general management of the office is concerned with providing a creative and supportive environment in which all members are able to operate effectively and efficiently. This includes providing easy access to reliable and current information and the use of office standards and masters. The management of individual projects is concerned with the accurate estimation of design effort and appropriate resourcing of projects to ensure that projects are delivered on time to an agreed quality standard. This includes the use of standard protocols to ensure a consistent level of service across the project portfolio, as well as encouragement and support from the design manager. Combined, these different levels of management should help specifiers to produce consistent standards of work.

Understanding how designers interact within the design office is paramount to the creation of excellent architecture and the profitability of the architectural business. By observing designers undertaking their daily work and encouraging feedback it is possible to implement and/or adjust managerial frameworks to better assist the designer in his or her task. Failure to understand the needs of the professionals working within the office may have an adverse effect on the ability of the office as a whole to perform. Similarly, understanding the relationships between independent projects as they pass through the office can greatly assist with the resourcing and coordination of design work.

The task of specifying is difficult to define since it continues, with varying degrees of frequency, throughout the design and construction phases. Many of the actions that the specifier goes through are, in the main, subtle and difficult to observe, a problem that is increasing with the use of computer based design systems. As a result, the process may be difficult to manage unless it is fully understood and the implications of decisions taken are recognized. Getting it wrong is expensive and, therefore, adequate systems need to be in place to prevent mistakes extending beyond the office boundary.

When things do go wrong, the tendency of senior managers is to blame the specifier. But before doing this it would be appropriate to consider the managerial structure within which the specifier works. First, the organization needs to employ staff who are competent and then allocate them to tasks appropriate to their levels of experience and knowledge. If inexperienced, then they should be adequately supervised. Second, a managerial system to prevent errors should be designed and implemented. Quality management systems and office quality-control checking procedures are essential tools, as are regularly maintained and updated office standards and masters. When mistakes are found, it is essential that the office procedures are checked and then specifier's actions audited for conformity with these procedures. If necessary, the system or systems can then be revised to prevent similar mistakes happening again.

The specifier's milieu

The innovativeness of the design office is related to the organizational culture of the office and its market orientation, or degree of specialization (Emmitt, 1999, 2007). Office size may also be a determining factor. The organizational culture, which is an integral part of the specifier's daily environment, will, to a greater or lesser extent, influence how they behave in the design office and will colour their decision making processes. This culture comprises the collective experience of the office, its management structure and working methods, possible use of standard details and master specifications, and the sources of information that it makes available. All this, as well as the previous experience of the specifier, will influence the specification process. So too, will the reliance on the personal and office palettes of favoured products and manufacturers (Fig. 6.1).

Managerial control

Regardless of size or market orientation, every design office needs to have someone in control, usually the senior partner or managing director of the office. In large organizations, the manager may be the design manager, someone who takes responsibility for design decisions made by his or her design team. These managers will influence the behaviour of individual specifiers through general policy decisions, individual project management and the day to day design office management. Control can be divided into three levels: policy decisions, individual project control and day to day managerial control.

- *Policy decisions*: policy determines how the office is managed and how the specification process is controlled. The use of quality management systems,

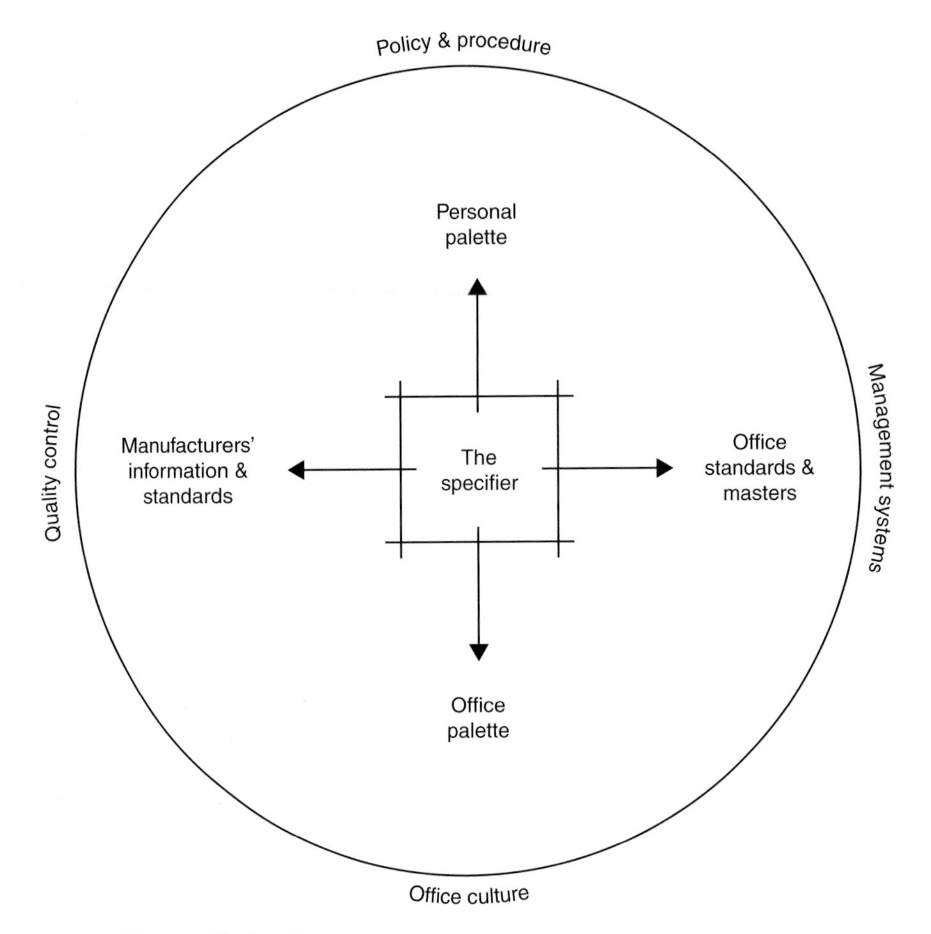

Fig. 6.1 *The specifier's milieu*

master specifications, preference for performance or prescriptive specifications, etc., will colour the behaviour of specifiers in the office.

- *Individual project control*: this will be tailored to suit individual clients and the characteristics of the design task. Quality parameters will (or should) be set out in the project quality plan and be reviewed at regular intervals via design reviews.
- *Day to day managerial control*: management of individuals within a design office varies widely, from leaving specifiers to make their own decisions with minimal input from their manager, to very tight control where decisions are closely monitored, and approval is required from the design (or technical) manager for the slightest variation in office procedure. Managers will also influence the process through their managerial style, be it autocratic or democratic.

Managerial control and support depend on clear leadership from managers and appropriate frameworks for undertaking the job. Related issues are the availability of design information to avoid guesswork and enabling informed decisions to be made, and the ability of the design office to learn from its collective experience. This involves making use of feedback opportunities to ensure continual learning. Designers need to be able to work together on and across projects, and the ability to communicate and share knowledge informally is necessary to avoid design errors and wasted time searching for information.

Design control

The amount of control an individual has over 'their' design project can be an emotive issue. A high degree of autonomy leads to a sense of ownership and pride in the job, while a low degree of control may result in individuals feeling helpless and undervalued. It is not uncommon for conflict to occur between the amount of individual control desired by designers and the level of control that is exerted by managers. Lack of personal control over their work, either perceived or real, caused by too much managerial interference may result in staff becoming less proactive. Individuals working in design offices tend to be extremely committed professionals, constantly striving for perfection. Ensuring a good fit between individual needs and organizational support will help specifiers to be self managing, making the job of the design manager considerably easier.

Control over the design itself is reflected in the amount of control the design or engineering office has over the specification process. This relates to the orientation of the professional office to the construction sector and the attitude of office members to the degree of design control (loose or tight). Combined, these factors influence the individual specifier's behaviour, from the choice of specification method, selection criteria and attitude to changing the specification after it has been written and issued to the contractor.

Quality control and assurance

For professional service firms, such as architects and engineers, quality control (QC) is concerned with checking documentation against predetermined standards. Checking drawings, specifications and associated documentation before issue and the checking of other consultants' documentation for consistency with the overall design concept will help to ensure the quality of the information provided to the builder. Quality control is also achieved by adherence to current codes, standards and regulations.

Quality assurance (QA) is a formally implemented management system that is certified, and constantly monitored, by an independent body, such as the British Standards Institute (BSI), to ensure compliance with the ISO 9000 series. Quality assurance is a managerial system that states what an organization will do (in documentation), doing so by defining set procedures, and proving that they have been carried out. The process is designed to give the customer a degree of confidence that the promised standard of service will be delivered. Quality management evolved from early work on QC in the American manufacturing industry, but it was the Japanese who developed the idea to what it is today. From the 1950s, they contributed to the Japanese revolution in continuous quality improvement, a revolution that has spread worldwide. Widely adopted in manufacturing, quality management systems have taken longer to gain widespread acceptance in the building industry, although many contractors and professionals now have certified quality management systems in place or claim to be working towards QA. In attempting to please clients through a total quality management (TQM) philosophy, a step by step approach to continuous improvement, known in Japan as *Kaizen*, TQM has gained widespread acceptance. It is a people focused management concept, a soft management tool, engendering pride in one's work and the desire to improve upon past success.

Office standards and masters

For the individual, the specification process is a research function from which decisions are made and communicated to others. To operate efficiently and effectively, the specifier needs access to relevant literature and to tools, such as the master specification, to help him or her to complete the task with the minimum of effort in the time available.

In the majority of design offices typical details and specification clauses are customized to suit the organization and hence become office 'standards' or 'masters' that are used to save time and ensure a degree of consistency. These standards are based on good practice (as viewed by the design office) and the collective experience of the office. Some details and specification clauses will be unique to a particular design office; others will be amalgamations of typical details, typical specification clauses and manufacturers' details and specifications. The common practice of using office standards is sometimes criticized for stifling innovative design, although it saves the specifier time and reduces the risk of failure. Office standards encourage good practice and save time, enabling the same drawing or specification to be used repeatedly.

Standard formats can be an effective tool in the quest for consistency of service provision and efficient use of staff time. Standard details and master specifications provide an excellent knowledge base from which to detail familiar buildings, and many organizations try to prevent employees who leave from taking such 'knowledge' with them to a competitor. They form part of the organization's collective knowledge and can be used by less experienced staff, as long as the process is monitored and checked by a more experienced member of the office. This information can be broken down into two categories: those products that are approved and those that are not. Effective use of such standards and masters offers a number of benefits, but there are also several pitfalls to be avoided.

Advantages

- *Quality control*: because standard details and specifications have been tried and tested by the design office over a number of years, they should be relatively error free. They will have evolved to suit changes in regulations and to accommodate feedback from site. They provide consistency where there is some turnover of staff and a pool of experience to guide younger, inexperienced staff. Because they are familiar, tried and tested, standard details represent an effective means of QC when applied correctly. Checked and updated at regular intervals, standards and masters may contribute to the quality management system of the office, reassuring clients and practice principals alike. This checking and updating must be a considered part of the management of the office and does involve some investment of time.
- *Time management*: the use of standard details and master specifications can save the design office time and money because common details and clauses do not need to be reworked, merely selected from the design organization's knowledge base. With increased downward pressure on professional fees, the use of office standards can help to ensure that the commission is profitable. Indeed, there may be little time available to investigate alternatives.
- *Risk management*: the use of tried and tested specification clauses helps to limit the organization's exposure to risk. Essentially, it is a conservative, or 'safe', approach to design.
- *Benchmarking*: when faced with an unusual detailing problem, the standards and masters form a convenient benchmark from which to develop the detail and help to evaluate its anticipated performance in the completed building.

Disadvantages

- *Perpetuated errors*: where errors exist in standards details and specifications (and they do), the errors are perpetuated through reuse on many projects until such time that the error manifests itself, sometimes after a long period.

Unless careful checking and updating are undertaken, the use of standards can prove a dangerous habit. Feedback from completed jobs (see below) is essential to this process.

- *Incorrect application*: inexperienced members of the design office are often left to apply standards with little or no supervision. There is then a real risk that they may apply details and specification clauses incorrectly, and managerial control is essential if costly errors are to be avoided. Auditing the specification process is important in tracing specification decisions and identifying areas for improvement.
- *Inertia*: as already suggested, the use of an office standard or master may inhibit the use of a better alternative when this is available.

Care must be taken in managing the use of standard products and details not to prevent the use of alternative solutions when conditions warrant. One of the authors was acting as a consultant to a large firm of consulting engineers when he was asked about the possible difficulty of doubling the load on cast iron columns. The firm had a standard detail for restraining the top of bay windows in rehabilitation work that involved the casting of a cranked concrete beam across the top. This presented no problem when the corners were masonry, but in this case they were cast iron. Although he was able to suggest a solution in steel that was liked by the engineer, he was told that it would not be approved because it was 'not the office's standard detail'. The office had a large proportion of young engineers with limited experience, and standard details appeared to be its way of coping with this.

The master specification

While a graphic representation of architectural details is used to control the form and appearance of the building project, written specifications are used to control the quality of the materials used and the quality of the workmanship. As the written specification controls quality, the development of office standard specifications is equally as important as that of standard details. These office standards are frequently linked to national and international standard specifications. As with standard details, standard specifications are used by organizations to save time, merely adjusting the master document to be project specific. Given that many design offices tend to specialize to a certain extent (e.g. housing, offices, hospitals), the development and maintenance of a carefully written master specification makes sense because of the major savings in time and the reduction in risk.

The master specification is essentially a library of specification clauses used by the design office on previous occasions that have been assessed for technical suitability. Filtered, coordinated and updated on a regular basis, it forms a vital part of the design organization's expert knowledge system. It is not to be confused with rolling specifications from job to job (discussed below). Regularly reviewed, the master specification can save individual specifiers considerable time and effort by reducing repetitive tasks. Correctly managed, over time, it will help to ensure consistency because all project specifications are drawn from it. It will maintain and improve quality through feedback of good and bad experiences, help to keep the cost of production down, and aid the coordination of information. Thus, the master specification is a crucial resource for helping to ensure QC and also providing a quality assured service to clients. The more effective and easier to use the master specification becomes, the greater the potential efficiencies and hence profit for the design organization. This source of information is a valuable resource and must be managed accordingly if it is to remain of use to specifiers in the office (Fig. 6.2).

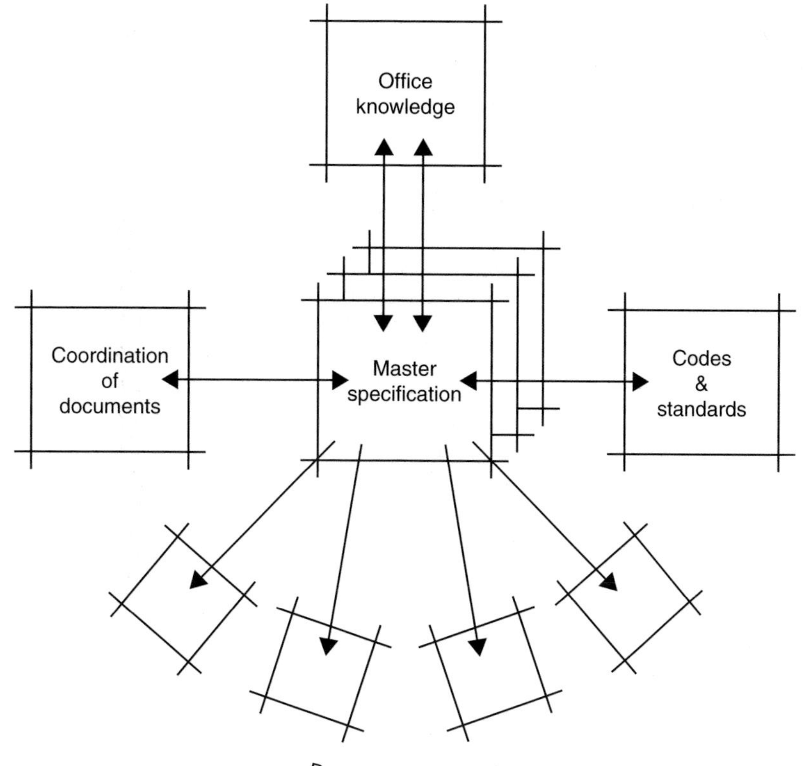

Fig. 6.2 *The master specification*

Controlling the master specification

As the master document needs to be maintained on a regular basis to retain currency, someone in the office must be responsible for its upkeep, i.e. responsible for its regular review and for ensuring that all alterations and changes are checked and recorded in accordance with the quality management system. In small offices, an individual will be doing this in addition to their other work and time must be properly allocated to the task. In medium sized offices, the task should be carried out by someone with responsibility for technical matters. In larger offices, the chief specification writers will spend all of their time on specification matters and may well have assistance from other technical staff to collect and analyse technical and product information. Because coordination is essential, it is common for design offices to set up the master specification based on a standard model, such as the National Building Specification (NBS), and stick to that. Thus, initial choice of commercial master is an important managerial decision.

A key to efficacy and quality is the ability of the individual in charge of the master specification to maintain it. Done well, all specifiers and their project specifications will benefit. Done badly, all specifiers and their project specifications will suffer. This means that an essential characteristic of the chief specification writer is the ability to keep up to date with new products and with new codes and standards. A greater challenge is to keep up to date with technical changes to existing products, with changes in methods of construction, and with revisions to standards and codes when they are issued. This person also has to have an awareness of contractual issues and legal liability. Clearly, such changes may necessitate revisions to the master specification, which must be the responsibility of the chief specification writer. All changes to the master specification should be recorded and their potential implications noted and communicated to the specifiers in the design office to avoid abortive work. Only the chief specification writer should be allowed to make changes to the master. Given the importance of the document, some design offices operate a double-checking system where someone other than the chief specification writer checks the document. Although time consuming, it is good office practice.

Project specifications

Individual projects will require their own bespoke project specifications. Just as every site is different from the last, so too are buildings and their specification. The quality of the written project specification will be determined by office policy and the abilities of the individual doing the writing. In offices where different designers write their own specifications, there may be a wide variety in quality unless they follow clear office guidelines. The use of quality management systems and reliance on the office master (and/or a national standard format) can help to make the specifications more consistent within the office.

Specifiers should be able to use a master specification as their starting point, in the confidence that it is up to date and free of errors. However, the degree to which individual projects draw on the office master specification will be determined by a project's characteristics and its compatibility with the master. Problems may be experienced in situations where a design office concentrates on, for example, residential work, and then acquires a commission for an industrial unit. In such situations, the master specification will be of less use than with the normal housing projects, and greater care is needed in its application to this unfamiliar situation. Decisions about product selection may be referred back to the design manager for approval. As noted in later chapters, great care (and adequate time) is required when dealing with products and information new to the specifier.

Rolling specifications from project to project

Rolling specifications are documents that have been used for a previous project and are simply rolled forward and adjusted to suit the next one. Their use is widespread but should be avoided because there is serious danger of including text that is inappropriate, and excluding that which should be included. Over time, other dangers such as references to superseded standards and discontinued products are real possibilities. This invariably leads to queries on site, over and/or under ordering of materials, additional costs and claims. It is a lazy and potentially hazardous approach to writing project specifications.

By its very nature, rolling a specification from one project to the next is 'convenient' when time and other resources are at a premium, i.e. when the specifier has not been allocated enough time to complete the task. Because the work is being rushed, problems occur largely because of inadequate checking of the text and inadequate checking against other project information such as detailed drawings and schedules. In the worst examples, site personnel become so frustrated with inconsistencies between the written specification and other project information that they simply stop reading the former. (One is tempted to ask: why bother writing one in the first place?) This leads to the dangers of changes on site and inappropriate levels of quality of materials, implications on cost and programme, and the enhanced risk of claims being made against the design office. Because of these problems, their use is not recommended, and well managed offices have managerial systems that prevent their use. Individual projects require individual project specifications, based on the collective knowledge of the design office, not on the idiosyncrasies of the specifier.

A systematic approach

Specifiers have different ways of working, but it is regarded as good practice to develop and build up the specification as the detail design proceeds. Once the first draft has been completed, it is then a case of editing the document to suit final design decisions. As with the master specification, someone other than the specification writer should check the project specification. Sometimes the design manager, sometimes the chief specification writer, does this job. It is poor managerial control to issue specifications and drawings without a comprehensive check for errors and coordination. If the master specification is kept up to date as changes to materials and codes occur, then there should be no need for feedback from individual specifications. However, good and bad experience of materials, products and working practices gained from individual jobs should be considered and the master document revised to accommodate new knowledge.

A systematic approach to project specification writing is recommended. Figure 6.3 illustrates the three main phases and the steps to be followed by the specifier. This starts with the overall strategy, followed by the writing procedure, and concludes with feedback. It is acknowledged that these steps may not necessarily be followed in a strict order, but the diagram should provide useful guidance.

Design reviews and coordination

No matter how good the members of the design team, and no matter how effective the QC and quality management system, discrepancies, errors and omissions do occur. Such errors are frequently related to time pressures and changes made on site without adequate thought for the consequences for other information. Many faults in buildings can be traced back to incomplete and inaccurate information and also the inability to use the information that has been provided. Discrepancies between drawings, specifications and bills of quantities can, and do, lead to conflict. Some of these can be avoided, but some slip through the net. Regardless of the sophistication of the technologies used to minimize mistakes and ensure coordination, it should be remembered that people make the decisions and input the information. Thus, errors may occur.

Checklists are a useful tool to aid coordination because they help the specifier to check that all the necessary information has been provided in the written specification (and highlight superfluous items). They also help to ensure that specified items are consistent with the drawings and schedules and that duplication is avoided. Offices tend to develop their own bespoke checking procedures that work for their particular way of doing things. This is particularly important when some of the production drawing packages have been outsourced because

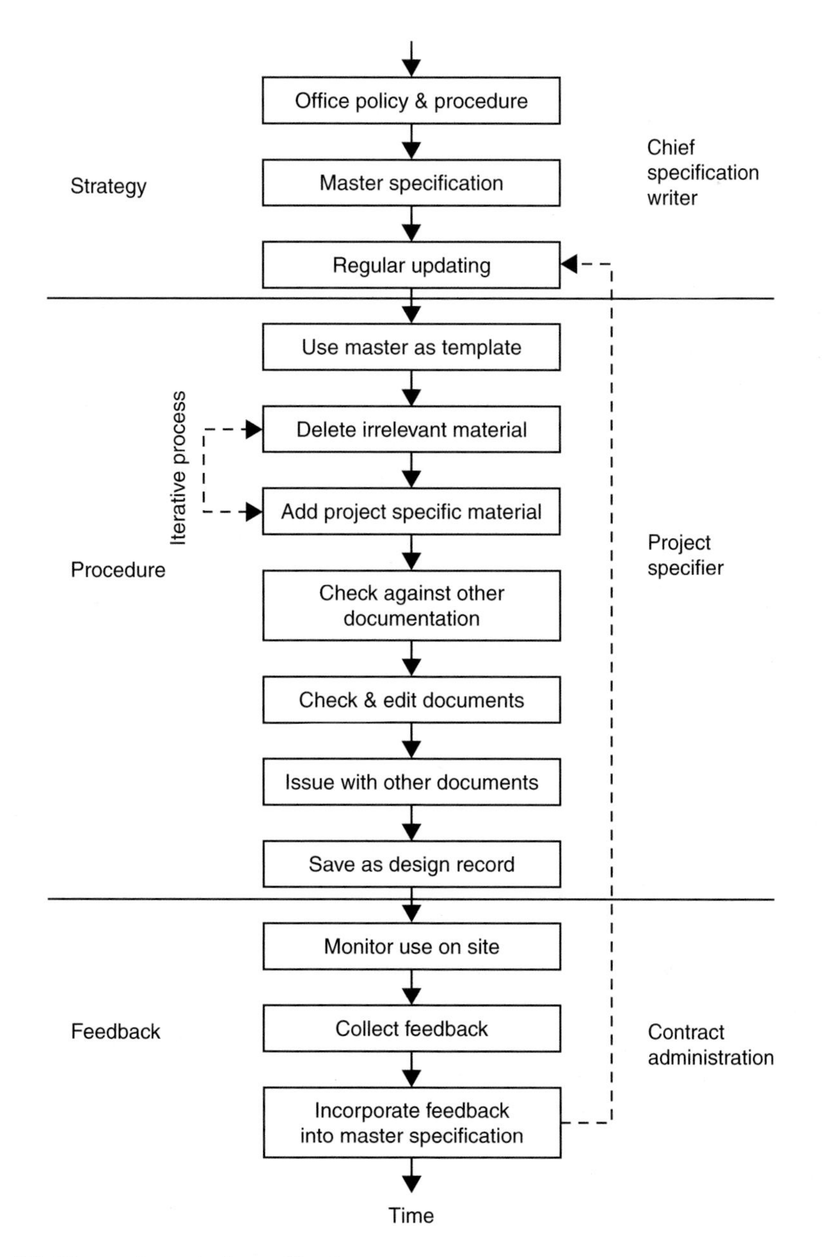

Fig. 6.3 *Stages in project specifications*

control at the design review then becomes more critical, regardless of the quality of the outsourced work.

Accommodating design changes

In many design offices, it is common practice to write the specification as the detail design stage proceeds. It is not uncommon for design changes to be made during this period, and so these changes need to be compared with the written specification and the latter revised, if necessary. Unfortunately, some designers fail to do this, and the information received on site often contains discrepancies between the details and the written specification.

In an ideal world, the process of producing the production information would be a smooth affair with everyone contributing their information on time, with the information received being complete, cogent, error free and sympathetic to other contributors' aims, objectives and constraints. In reality, this is rarely the case, regardless of the quality of the managerial systems and the effectiveness of information coordination. Project teams are often assembled for one job, and the participants may not have worked together previously, and it is only towards the end of the project that teams start to communicate effectively. In the meantime, there is the potential for errors to occur simply because no empathy has been achieved. Because of this, the possibility of design changes occurring needs to be allowed for in any programming.

Of course, changes to the design can come from a variety of sources and not simply be generated internally within the design team. They can come from the client or from the contractor if the latter is involved early in the process. All changes need to be approved by the client before they are implemented, costed, their consequences fully considered and the change recorded.

Changes before construction begins

The written specification will form part of the production information that is issued to the contractor for competitive tendering or negotiation purposes. The contractor's estimators will read the specification alongside the bills of quantities and the drawings in order to arrive at a contract sum. With some types of contract, such as negotiated contracts, the contractor will be encouraged to suggest alternatives with a view to reducing the whole life cost. With competitive tendering it is common practice for the contractor to tender two bids, one based on the work as specified and the other based on alternative (cheaper) products. With both approaches it is important that the design team check the characteristics of the proposed alternatives before work commences, and that all agreed and approved changes are recorded.

Feedback

One of the most common complaints levelled at designers, especially architects, is their inability to return to projects to gather feedback. They seem reluctant to analyse projects to see what was successful and what could have been better, with resultant information helping to inform future projects. Where the office uses a master specification, the incorporation of feedback is essential for ensuring that the documentation retains its currency, i.e. by incorporating relevant changes into the master specification as soon as possible. This needs to be done systematically and will be an essential part of any QA systems implemented by the design organization. Another, more holistic, approach to feedback is the ability to reflect during the process and learn from it (discussed in Chapter 11).

Value management

With increased emphasis on cost and value for money comes a concern about 'over specification'. Given the cost associated with building components, product specification naturally forms a major part of any value management exercise. Under specification will become evident when a component or system fails, but over specification may only be evident when a building is finally demolished or substantially remodelled. Consideration of service life at briefing and subsequent design and specification stages can help to eliminate over specification, as can the use of value management techniques.

Structured value management exercises are concerned with exploring and identifying value to the client and then linking these requirements to the most cost effective (in terms of whole life costs) design solution. It is becoming common for value engineering/management exercises to be conducted at strategic stages in the development of the design. It is also becoming common to link this exercise to risk management techniques, the argument being that these activities are complementary (Dallas, 2006). The intention is to have a critical review of design solutions and product specifications before they are issued to the contractor, helping to maximize the value of the design by identifying and then eliminating waste. Thus, over and under specification should not occur.

Value management exercises are usually conducted as a facilitated workshop involving the main project participants. Sometimes consultants from outside the project team are brought in to look at the project from a more detached perspective. Various qualitative and quantitative tools are used to help to evaluate the design and the specified products in terms of the value they provide. The outcome of the exercise may be that alternative products representing better value are substituted for those initially specified. Therefore, specified products may be changed before the contract information is issued to the contractor.

Contractors also use value management tools both during the tendering process and the realization phase, with a similar objective to that of the design team.

Design control during the realization phase

Quality control (QC) is a managerial tool that ensures that work conforms to predetermined performance specifications. For manufacturers, with long runs of products in a controlled and stable environment, QC is relatively simple. Assuming that the technology and personnel are correctly deployed, it is easy to maintain a constant standard. When it comes to achieving QC on a building site, the circumstances are different. First, much of the work is carried out without the benefit of shelter from the weather, and emphasis on programming work to achieve a weather proof envelope as early as possible in the assembly process is a prime concern. Second, the number of different operatives present on site at any one time (sub contractors and sub sub-contractors) makes the monitoring of quality particularly time consuming for both the construction manager and the clerk of works. The levels of skill and experience of site personnel are often uncertain and work can be completed and covered up without anyone other than those responsible for its building knowing whether or not it complies with the standards set out in the specification.

To combat this, there have been various initiatives to move as much production as possible to the factory by way of prefabrication, leaving only the assembly to be done on site, which have been tried over the years with varying degrees of success. The effect has not always been an improvement in the quality of construction.

It is important to recognize that such changes involve an implicit change in the structure of the building industry and that the management of the professional service provided needs to adapt to this. The extensive use of precast concrete in the 1960s led to failures as a result of inappropriate materials, sometimes in direct contravention of the specification. This can be put down to a lack of an appropriate QC mechanism. The failures associated with the introduction of trussed rafter roofs can, in part, be traced to a failure to recognize properly the degree of professional scrutiny required; but poor site supervision was also a contributory factor (Yeomans, 1988b). The trussed rafter example also demonstrates that site supervision is vital where the operatives on site may either be unfamiliar with a construction process or not understand the significance of the details. This can literally be a matter of life and death. It became clear that many building collapses in the Turkish earthquake of 1999 were the result of inadequate reinforcement. What those actually making up the reinforcement seem not to understand is the vital need for adequate shear reinforcement in columns.

The degree of site supervision associated with any building process is therefore an aspect of QC that should be considered at the specification stage. If what one is specifying requires a degree of care and skill in its installation, one needs to be assured that that is available on the site. In some cases details may be designed simply to avoid possible errors. For example, if the calculations show that different sizes of screws are required in an assembly it may be prudent to specify the larger size throughout to avoid possible errors on site, in effect dispensing with the supervision that may otherwise be needed.

Using the written specification on site

If the written specification has been compiled and written in a professional manner one might hope that things would progress smoothly during the realization phase. Unfortunately, such optimism is often misplaced, since there is plenty of anecdotal evidence and a small amount of research indicating that the written specification is not always referred to. This may be because the written specification is not (or is not perceived to be) as thorough as it should be, or simply that the habit of site workers is to ignore it, and so ignore the quality of the workmanship required. Perhaps some of these problems are rooted in the failure to explain clearly the importance of the written specification in educational programmes, although the authors are unable to find any research that has attempted to address this. Whatever the reasons, it is crucial that the written specification is used by those doing the work as well as those responsible for managing construction.

The contractor needs to be confident that the written specification is complete and does not contradict information contained elsewhere, i.e. in the drawings, schedules and bills of quantities. Although some of the discrepancies will have been identified and dealt with during the tendering phase, it is not until the contractor starts to order materials and products that some of the less obvious discrepancies may become evident. This will inevitably result in requests for clarification and additional information, and often a claim from the contractor for extra monies.

Ensuring adequate standards of workmanship on the site is the responsibility of the contractor, who will do this largely through the selection and supervision of adequately trained personnel. It is not the task of the specification to substitute for this. One would hardly describe how to knock in a nail, but clearly there are situations where the workmanship required is, in some way, out of the ordinary and it is in these circumstances that some attention needs to be paid to this within the specification. However, the specifier needs to be aware of the limitations of this. It seems that the written specification is not read frequently enough by site operatives; nor is it understood sufficiently, a situation exacerbated by the use of temporary and often poorly trained operatives. Operatives will rely, for the most part, entirely on drawings and not look at a written specification unless instructed to do so by the contractor's agent. Even if they do read the specification, it is unlikely that any standards referred to will be available on site, nor are they necessarily familiar with the common used standards. In an Australian study, it was found that bricklayers were not familiar with the Australian masonry code and that their work (perhaps not surprisingly) did not comply with it despite reference to it in the written specification (Nawar and Zourtos, 1994).

The experience of the authors working in the UK is not dissimilar. On being called out to site to discuss a problem with the standard of workmanship, it was found that not only did the workers (sub contracted labour) not have a copy of the written specification, but also the site manager had not read the documentation. This became evident when a copy of it (and a package of detailed drawings) was found unopened in the site office. The project had been running for just over four months, and perhaps not surprisingly was experiencing more than its fair share of problems. Unfortunately, similar stories are not uncommon and a great deal of time and money can be wasted when the information provided is not used. This is an issue for contractors and their construction managers to address, and something largely outside the scope of this book. However, the design team should take care to explain the importance of the written specification at the pre contract meeting and remain vigilant when visiting the construction works. The appointment of a client's clerk of the works to check that the work is conducted to the required quality as set out in the written specification may be a sensible investment.

For those working on the site the challenge is to be familiar with, and have ready access to, relevant standards and codes of practice. In an ideal world we would be able to rely upon trained operatives, something that the apprenticeship system once gave us. But even if that was an ideal for so called traditional construction, it is unrealistic for a world of new products requiring new skills. One approach is to ensure that sub contractors are certified by the manufacturer of the product they are installing, especially when that product is not easily inspected once incorporated into the building. The range of factors on which we depend to obtain the required quality of workmanship varies from particular instructions contained within the written specification to the simple reliance upon adequately trained site operatives.

For some special classes of work the employment of specialist sub contractors operating under a certifications scheme that (supposedly) ensures the training and supervision of their operatives provides some assurance of workmanship standards. However, difficulties can still occur. On a contract involving structural carpentry it was found impossible to obtain timber of the required size and below the maximum moisture content specified. The only option was to obtain boards at the required moisture content and laminate the sections. Although the contractors were known for a good quality of structural carpentry this was an operation of which they had no previous experience. The workmanship was specified in BS 6446 and, in an attempt to ensure that standards were met, the engineer specified that boards were to be sanded before gluing and that samples were to be made for testing. The data provided in the test reports were to include the temperature during gluing. All this was done, but there were still uncomfortably large variations in the test results and the engineer at first suspected that the sanding had been omitted. However, on visiting the workshop it became apparent that while the carpenters had been aware of the need for the gluing to be done at a particular temperature they had not read the glue manufacturer's specification closely enough to realize that this temperature had to be maintained during the curing period. In Germany, the source of the glue, those carrying out these operations need to carry a certificate to do so.

Substitution of proprietary products

An area of interest to manufacturers and specifiers alike is specification substitution or 'switch selling', the substitution of a different brand product from that originally specified. The problem of specification substitution has been highlighted in the technical press (e.g. Hutchinson, 1993) and by manufacturers keen to discourage such practices (Hutchinson, 1995), although, with the exception of research by the Barbour Index, there is little evidence to quantify the scale of the problem. Reasons for suggesting alternative products tend to be that it is to overcome a problem with availability, to resolve a buildability problem that has arisen unexpectedly on site, or to save money for the client (this claim needs careful consideration). The real reason may be simply that the contractor is not familiar with the product and does not know where to obtain it (i.e. it is not stocked by the builders' merchant used by the contractor). As the case study (see below) has shown, it is worth checking the contractor's claims about the proposed substitution because they may not always be truthful. Less reputable contractors propose alternatives simply because they can make more money on the contract through cost savings that are not passed onto those funding the project. A particular trick of contractors and sub contractors is to wait until the last possible moment to request the change to try to put the specifier under pressure to make a quick decision. Some designers make snap decisions over the telephone and live to regret it; others refer the contractor back to the contract clauses and their QA procedures and will not make a decision without the client's consent.

Common tactics used by contractors and sub contractors are:

- Deliberately delaying a request to change a specified product until it is urgent, putting the specifier/contract administrator under pressure. The

proper response is to refer the contractor to the contract conditions, which would normally require the contractor to put a written request to the specifier, along with appropriate information, and allow the specifier time to make an informed decision. This usually stops the practice.
- Accidentally delaying the request, once again putting the specifier in a difficult position. The problem here is that there is no way of knowing whether the contractor is acting honestly. If in doubt, the response should be as described above.

It is not unusual for a contractor to propose alternative products from those specified by brand name under traditional forms of contract. In such situations, the designer (or contract administrator) has to be certain that the substitution proposed is of equivalent quality to that originally selected because specification decisions remain the responsibility of the design office. The original decision was often taken after an exhaustive search of similar products, so that specifiers are often reluctant to approve a contractor's request because of the time it then takes to check the performance characteristics of the proposed alternative. In situations where the contract administrator is not the specifier, e.g. a project manager, the request should be referred back to the specifier for an accurate evaluation; it is unlikely that the project manager has sufficient technical knowledge to make a decision to substitute.

Mackinder's sample of architects had two distinct views about changing specified products when a request was made by the contractor: either they were prepared to change to an alternative product if it was of equal standard, or they refused to change the named product (Mackinder, 1980). Hutchinson's advice (Hutchinson, 1993) is to stick to the original specification at all costs because the architect's office is legally responsible for any changes made to the specification regardless of who makes the suggestion (Cornes, 1983). The Barbour Report (1993) found that in fifty six contracts examined, just over half had experienced a change of brand name product after specification. Further evidence of specification substitution was provided in the Barbour Report of 1994, in which main contractors stated that they substituted alternative products to those originally specified in 10 per cent of specifications prepared by the design team, and sub contractors said that they, on average, altered 23 per cent of product specifications. Cost was the main reason reported for changing product specifications on the building site (Barbour Index, 1994).

Surreptitious substitution

Contractors and sub contractors have admitted to surreptitious substitution of products, sometimes referred to as 'breaking' the specification. The Barbour

Index (1993) found that contractors claimed to be making changes without the knowledge of the contract administrator and that sub contractors were making changes without the knowledge of the main contractor. Motivation for such action is financial gain, and because of this, it is difficult to obtain quantitative evidence of the true extent of such action. Another reason why people are reluctant to discuss the extent of specification substitution is that it is an act of fraud; clients are paying for specified products and getting something else. In conversation with construction managers for the purposes of this book, the extent of specification substitution would appear to be more common than has been reported. It is an area that deserves the attention of some research because it can also have a serious effect on the quality and durability of the building.

Both authors have experienced attempts by site workers to make surreptitious changes to specified products, although it is not possible to say how prevalent these practices are. In one example the author was called to site to resolve a quality problem with some facing brickwork and specially designed brick panels. On arriving at the site it was evident that the quality of the bricks was not acceptable and so the author telephoned the manufacturer to ask for an explanation. The response from the manufacturer was a little unexpected; they simply stated it was not their problem since the bricks on the site were not theirs. They had not received an order from the contractor. This resulted in the contractor being instructed to remove all of the work and replace it with the specified bricks at their own expense. It also resulted in the author losing faith in the contractor and the contractor being removed from the design organization's list of preferred contractors and suppliers.

In situations where the material is used externally, the town planning authority quite rightly takes particular interest; even so, there are many examples of contractors changing products in order to save money that have backfired because of this. In an example taken from a design and build project, a twelve storey office building in a prominent city centre location, the contractor changed the cladding to the external columns from stone to brickwork in order to save money because the contract was running over budget. This considerable cost-saving decision backfired when the planners served an enforcement notice to comply with the town planning consent.

That manufacturers are concerned about such actions is understandable. Manufacturers invest in product development and marketing. They also spend time on getting the specification through the action of their trade representatives. Once their product is specified, they may then have to ensure that it remains so until it is built in on site, as highlighted in Chapter 10. Substitution is where the manufacturers of cheaper products obtain many of their sales, and they will put considerable pressure on contractors and specifiers to make the change they desire. Contract administrators and clerks of works must remain alert to such practices.

Change of mind

It is not all one way traffic. Designers and clients have been known to change their minds, sometimes before a job gets to the construction phase, and sometimes when the job is in progress. Care needs to be taken that the change does not repudiate the contract. The contractor will probably want some form of compensation, usually in an extension of the contract period or costs for accelerating the work to accommodate the change if it is anything other than a very minor variation. The time between the initial specification decision and starting work on site may also affect the likelihood of changes being made. In a fast-track project, for example, the time span from planning permission to work beginning on site can be short. In contrast, where the time from specification to actual assembly of the product has been lengthy, some products might be changed by manufacturers during this time, circumstances can change and costs can also change dramatically, especially in periods of over or under supply. In these circumstances, changes may be necessary. On rehabilitation, repair and alteration work changes will be necessitated because of the nature of the work. Changes are inevitable, but they can be minimized if the specification is based on thorough investigative before work is begun.

Auditing the specification process

In line with a well implemented and managed QA scheme is a need to monitor the process, both in the design office and during the contract stage, to ensure compliance with the specification. In an attempt to control specification substitution, some manufacturers and researchers are looking at the possibility of barcoding their products and/or adding radio frequency identification (RFID) tracking devices for ease of identification during building and at any future date when the building may be remodelled and/or recycled and the materials recovered. Such schemes are also designed to prevent unlawful specification substitution, since it is possible to identify the exact location of products.

Resource allocations

Given the importance of specifying and specification writing, it is surprising how little time is allocated to it in many design offices. In the less well managed offices, earlier stages might have exceeded their time allocation so that later stages, perhaps seen as less creative, have to be squeezed if the office is to deliver the product information on time. The production drawings, and especially the specification writing, being at the end of the process, are the two stages that frequently lose allocated time, resulting in rushed work that is inadequately thought through or checked. The result can thus be a document with too many omissions and errors that inevitably provide the contractor with opportunities for claims and/or inadequate work. Time and cost are closely related, and the manner in which these two valuable resources are managed will affect the quality of the service provision and that of the finished building. Clients want a quality building for as little financial outlay as possible, and (of course) they want it delivered in a short period. From the designers' perspective, the budget is never quite generous enough to allow good quality materials, and the time frame to achieve a good design is always too tight. Builders are then on the receiving end of cost cutting exercises and tight programmes.

With careful planning and good managerial control, the majority of projects are delivered on time and within budget, but when things do go wrong it invariably leads to the need for additional time and/or additional expense. It may be true that one or two weeks in the design and construction stages is negligible in the overall life of a building, which may be 100 years. But this does not help the client who may be paying interest on loans, and this argument should be made at the briefing stage, not when the project is starting to run behind schedule. The point is that good management begins with the correct estimation of the resources and time required at the beginning of the work and the discussion of this with the client at the briefing stage.

As discussed above, when allocating resources for a project, the design manager must allow adequate time for writing and checking the specification before issue. The task of specification writing should be clearly separated from the task of producing the production drawings. In practice, the tasks of detailing and specifying are often difficult to separate, but they are quite different tasks and must be costed accordingly. To produce a set of comprehensive, error free drawings takes time, and so does the writing of a comprehensive, error free specification. They are interrelated, but separate, tasks and must be resourced accordingly even when draft specification clauses are written as the design proceeds.

Resource allocation takes on an even greater importance when dealing with existing buildings. Even where extensive investigations have been carried out, it is unlikely that precise requirements can be established until the building is opened up, i.e. as the work proceeds. Allowance will have to be made for changes to the specification (covered by the contingency sum), and the design manager must allocate sufficient time for the specifier to deal with such changes (covered by some contingency in the programme). This is important, because the time pressures placed on the specifier to make a decision may be more critical than on new build projects. The authors' experience of repair and rehabilitation projects is that clients are often reluctant to allow sufficient opening up of the building until the last moment, primarily because they wish to keep the building in use (health and safety concerns permitting) while the work proceeds.

In conservation work there is the added difficulty that clients simply fail to realize that money and time spent on the initial investigation will most frequently save both in the long run, and it may be difficult to persuade them of this. What they are painfully aware of is that fees are being incurred when they see survey work in progress while nothing else is happening on site. Their fears may be assuaged by ensuring that they receive a report on this work. The preparation of a thorough report of the investigation is of value to the designer, ensuring that the implications of the findings have been dealt with, and to point out any additional investigation needed during the execution of the work. It is also valuable to demonstrate to the client that money on it has been well spent and again to indicate where there are still uncertainties to be resolved. Such reports need to be carefully written and well presented, and adequate time for this must be allowed.

Staffing requirements

In an attempt to be cost effective, and hence competitive in the market for professional services, there is a tendency to use the cheapest available person for specifying. However, the cheapest person available does not necessarily mean the cheapest final project because staff with lower hourly rates tends to be those who are less experienced and, it may be argued, not suited to writing specifications. From a manager's perspective, it may be useful to consider staff in terms of their experience, rather than their cost per hour (which may not necessarily be comparable). Three kinds of staff may be identified:

- *Inexperienced staff*: these are usually students or the recently qualified that are the cheapest resource in staffing terms. However, the need for constant nurturing and supervision makes the true cost of this resource considerably higher than it may appear from a balance sheet. A considered mix of advice from experienced colleagues combined with an ability to question conventional wisdom is desirable.
- *Experienced staff*: experienced staff are a design organization's greatest asset. Capable of working with minimal supervision within the office managerial system, they can often produce accurate work fairly rapidly.
- *'Over experienced' or complacent staff*: care should be taken to ensure that experienced staff stay up to date with current developments and do not rely entirely on over familiar (and rarely challenged) solutions. Reallocation of duties usually dispels any complacency very quickly.

What usually happens in a design office is that the design manager has to use the staff available at the time (those who are least busy), rather than those best suited to the job. This can be avoided if the office is managed using the sequential system (Sharp, 1991; Emmitt, 1999, 2007) where the job is passed along the supply chain, a systematic approach that can be very cost effective if well managed.

Time and professional fees

Regardless of whether information is provided on paper or in digital form, both time and other resources are required to complete the task in a professional manner. Time is required to research possible solutions, to think about the consequences of design decisions, to produce and check the drawings and schedules, and to coordinate these with each other. Time is also required for other consultants to integrate information with their own. Time will also be needed to make changes, because there will be some. Apart from all that, time is also needed to record and manage the process. All of this must be reflected in the level of professional fee income.

Time is the most precious resource and the one that no one ever appears to have enough of. No matter what the task, we would all like longer to complete it (or do it better): there is nothing unusual in this. Time has an economic value, and for commercial concerns, the sooner clients receive their building the greater the financial return. Similarly, building designers and builders able to minimize the amount of time required to assemble a building, from inception to occupancy by the client, have a competitive advantage over those who cannot: a service many clients are willing to pay a premium for. To do this requires extensive knowledge of design, manufacturing and assembly, as well as managerial skills.

Among other factors, the effective management of the specification process is fundamental to ensuring a quality service and a quality product in the agreed time scale and for the agreed fees. This is an area that authors of books on office management and specification writing have avoided, but it is covered particularly well in Chapter 3 of the NATSPEC guide. Here, Gelder (1995) suggests that the time required to write specifications varies from 15 per cent for smaller projects down to around 2.5 per cent for the largest. No evidence is provided to back up these figures, and the author acknowledges that the figures are a 'crude' guide. However, if correct, the implications are that the smaller offices that are handling such small projects have the most to gain by a careful consideration of how they might improve the efficiency of this process. The actual time taken depends on a number of factors, including access to available sources, expertise of the writer, thoroughness, timeliness and the quality of the input from others party to the process. The NATSPEC guide does go on to provide some additional guidance, but also adds that programming the specification is 'not an exact science'. The present authors beg to differ.

It may be an obvious statement, but the time taken, and hence the effective programming of the specification process, is influenced by the manner in which the design organization is managed, and this varies widely from the exemplary to the chaotic (Emmitt, 1999). Simply because it is a process particular to a specific design office, it is necessary for the design manager to set targets and monitor the time taken so that future programmes can be planned with more accuracy. Use of data collected on time sheets, feedback meetings and monitoring can provide the information to allow some very accurate planning and improved quality of work. This holds true for new build and work to existing buildings.

The cost of producing information is often underestimated and is not particularly well controlled in many design practices. Given the quantity of drawings and associated documentation that have to be produced during the detail design stage, the careful management of their production and especially of the time spent in producing them is critical to the profitability of individual jobs and will influence the long term viability of the business. Each and every drawing, schedule and specification should be costed as a percentage of the job. Allowances should also be made for unforeseen design changes and dealing with requests from the contractor for changes to the specification, which can easily affect a job's profitability.

Outsourcing

Organizations have been quick to realize the potential cost savings and increased organizational flexibility afforded by outsourcing their non core services. Some design offices have outsourced aspects of their work for a long time, for example specific detailing requirements to consultants with whom they have developed informal working relationships. Indeed, many design offices rely heavily on contract staff to help in busy times.

Outsourcing packages of work to other professionals can form an effective way of managing the design organization. Some design practices are starting to specialize in design and information management, i.e. they do the conceptual design work but outsource the task of producing the project documentation to a variety of specialists, ranging from technically orientated professional design organizations to specialist sub contractors and suppliers. This is similar to the French system, where detailed design work is carried out by the *bureau d'études*.

Some organizations maintain a master specification (as a control over standards) but outsource the project specification for particular jobs. Whether or not the specification writing is one of the core services provided by a design office clearly depends on the market orientation of a particular office. However, given the importance of the master specification, caution needs to be exercised if this service is to be outsourced because the office will effectively lose control of the knowledge and the investment contained in the master specification.

Digital specifications

With the growth of computer usage, increase in computer power and more sophisticated software, the potential for managing the specification process digitally has become a reality for even the smallest design office. Standard specifications, such as the NBS, and bespoke office standards are available and are widely used. These electronic formats tend to follow the same layout as their paper based forerunners, but have the advantage of being able to import information from other sources quickly, from either the office master specification or manufacturers' product specific information. Software packages offer the added benefit of being updated on a regular basis (assuming that subscription is maintained), and thus the possibility of using outdated clauses is minimized. Hardware capabilities aside, there are two essential requirements: first, ease of use for the specifier, and secondly, compatibility with other documentation.

- *Ease of use*: efficiency can be increased through the use of specification writing software, but only if it is simple and quick to use. Specification writing involves the transfer of information, and the easier this is to import, cut and paste within the document, the better. Search facilities are also vital to find particular words and/or clauses to help the specification writer to complete the task expediently.
- *Compatibility*: because the specification is such a central document within the overall project information, it is essential that the software is compatible with other software used by the design office. It should also be compatible with software used by other participants in the design process so that information transfer and hence coordination can occur freely. Typical information sources are drawing files, product data library, manufacturers' technical information, current standards and codes, and the bill of quantities/schedule of works. The project specification must be based on the same software as that for the master specification.

As with a paper based system, the digital files must be clearly labelled and dated to avoid any confusion. When upgrading hardware and/or software, care is needed to ensure that information (and hence organizational knowledge) can still be accessed quickly and easily.

End of chapter exercises

- Discuss the merits and demerits of outsourcing the specification writing to:
  (a) a professional office located in the same street as your office
  (b) a professional office located in a different country from your office.
  What are the implications for managing the specification work in the office?

- During a value management exercise a number of specified products have been identified as offering poor value for the client. However, these products were identified by proprietary name in the client's specification. How do you proceed?
- On a routine visit to the construction site you discover that a section of work is not to the standard as stated in the written specification. What do you do?
- You have developed a good relationship with a particular manufacturer during the course of a project. In particular, this manufacturer offered a lot of technical advice and responded to problems quickly. You wish to use this manufacturer's products on your next project; however, the client has insisted on using performance specifying.
  (a) How could you ensure that this manufacturer's products are used?
  (b) How do you share your positive experience of this manufacturer with your colleagues?

Specifying 'new' building products

So far, this book has been dealing with theoretical and practical considerations relating to the specification of buildings. This chapter marks a change in emphasis. The remainder of the book will focus on how specifiers behave in practice, with particular emphasis on the uptake of building products that are 'new' to the recipient.

Approximately 600 patents relating to building and civil engineering are granted in the UK every year, but only 5 per cent of these reach production, providing the building industry with around thirty new patented products each year. In addition to these 'new' products, there are numerous minor product improvements that are introduced by manufacturers to prolong a product's life in the marketplace. Added to this, building is becoming more international, so that there are products that are developed in other countries and introduced to the British building industry and these present particular issues that need to be addressed.

Getting a new product adopted is not a simple process. New products and product improvements are dependent on decision makers in the building industry for their selection, and either ignorance of them or conservative behaviour by specifiers tends to favour established products. How, then, are these products adopted? What processes are involved in the selection of a new product by a specifier? The office manager or contract administrator will need to be aware of this process if new products are to be specified. It is also in the manufacturers' interest to understand how specifiers make their selections if they are to market their products effectively.

New ideas and products

The focus of work on the adoption of new ideas in architecture and building design has been the province of the architectural and the economic historian, both of whom have taken a broad view of the way in which new ideas have been adopted. Studies of medieval timber framing in England have been concerned with explaining the dissemination of the quite different regional traditions of carpentry, a process that has been seen by different scholars as both a geographical and a social process (e.g. Mercer, 1975). While the former simply depended on the gradual spread of knowledge of different forms among carpenters, the latter assumes that increasing wealth allowed those lower on the social scale to aspire to the standards of their betters during a period when the timber frame was exposed. It was therefore an aesthetic consideration that drove development. The significant difference between these is that the former implies simple technical improvements adopted by the carpenters, while the latter requires that the new forms be demanded by the customers.

Yeomans (1992: 144) has noted that the spread of new structural forms during the late seventeenth and eighteenth centuries was influenced by the dissemination of knowledge via peer group contact, by copying and by knowledge gained from illustrated carpenters' manuals, i.e. several mechanisms were involved, mechanisms that have also been observed by others. Peters (1988), who examined rural buildings in a small region of England, suggested that the traditions of carpentry were influenced by the spread of building books during the eighteenth century. Benes (1978), in a study of the diffusion of aesthetic ideas in rural New England meeting houses, used the term 'diffusion' to advance a theory for the spread of a building style across a landscape. Although he made no mention of diffusion models, the approach is consistent with the spatial diffusion models (e.g. Brown, 1981). Benes noted that new ideas spread into rural areas from the urban centres through knowledge gained from design books, travellers' accounts and newspapers. Furthermore, architectural fashion played a part, with new meeting houses being designed to look like, or be an improvement on, those in neighbouring towns. In some cases, the same builder was employed to achieve this objective. A particular architectural fashion, and its subsequent diffusion identified by Benes, was the first use of coloured paint on the meeting houses.

The concern of the architectural historian has been to identify and explain the origins of new ideas, normally focusing on particular designers, but there has been little work that has looked at the process in general. While Yeomans (1992) was concerned with the spread of a structural idea and did make reference to the 1962 diffusion ideas of Rogers, a more recent study by the same author (Yeomans, 1996) considered the way in which theoretical studies influenced concrete mix design and made greater use of the Rogers diffusion model. In particular, this noted the influence of the leading journal in promoting the innovation to potential adopters. However, this recognized that the application of a general diffusion model to the building industry was complex because the social structure of the building industry was more complex than those involved in the studies reported by Rogers.

Innovation

Innovation has been important to economic historians, of whom Bowley has undertaken the most comprehensive studies of the building industry. Her

Innovations in Building Materials: An Economic Study (Bowley, 1960) reviewed a wide range of products, while *The British Building Industry: Four Studies in Response and Resistance to Change* (Bowley, 1966) considered a small number of innovations in some depth and is the more significant in the context of this study. These provide a valuable insight into the adoption of building product innovations, although they were not conceived as diffusion studies. Most of the innovations that she discussed in the first of these were associated with major building materials, in particular advances in their production and their associated cost benefits. She concluded that innovation in building materials is influenced by the desire of the manufacturers to hold and extend their markets, rather than in response to particular needs, i.e. innovation in building products is influenced by market push rather than demand pull.

In looking at the introduction of a small product, the concrete roofing tile, Bowley (1960) considered the way in which its producers manipulated the structure of the industry. The Aisher family began by selling their concrete tiles as a product, but found that the cost savings were being used by builders to increase their profits, rather than being passed on to customers. This failed to realize the potential sales, and in 1926, they set up as roofing sub contractors. The success of this operation led to the establishment of the Marley Tile (Holding) Company in 1934. The implications, in the context of this study, are that an adopter may be buying into a change in the structure of the industry. Another example of a product adoption being accompanied by a change in the structure of the industry occurred with the adoption of the trussed rafter roof in Britain in the post war period (Yeomans, 1988b). In the first case, a manufacturer became a subcontractor, and in the second case, materials suppliers became manufacturers.

An important aspect of the adoption of the trussed rafter roof in Britain was that this product had its origins in the USA, where roofing practices were quite different. The failure to recognize both the significance of these differences and the changes that had been occurring in British building practice in the preceding decades resulted in some initial failures. Although conservative behaviour in the industry can be a barrier to the adoption of an innovation, this was an example of rather too enthusiastic adoption of a product, partly because of ill informed sales methods (Yeomans, 1988b). This is discussed in further detail later.

Bowley's second book considers major changes, such as the introduction of reinforced concrete as a framing material or the new methods for the design of steel frames, rather than the introduction of small products. Such developments require a major restructuring of the industry, or of the way in which people design. This is an interesting issue in itself, but one that can only be explored as a historical phenomenon. The significance of this work is that it examines innovations that were largely rejected by the industry and considers why this should have been so. Reinforced concrete failed to achieve the market share that it might have against steel frames that were established first, and engineers resisted changes in design methods proposed by the Steel Structures Research Committee.

Other economists, such as Stone (1966), have looked at the relationship between innovation and the cost of labour and materials and found that innovation in building has been generated in different ways: clients have set new problems, designers have used new materials to solve problems and contractors have used new materials to reduce the cost of construction. All these have led to an increased range of materials and an increased number of possible methods of building, suggesting a demand pull, and apparently contradicting the views of Bowley.

The modification of existing buildings has also been an area for the adoption of innovations. In Britain, the adoption of plastic based window frames was associated with the retrofitting of double glazing in houses. The adoption of solar heating by households in a California neighbourhood was studied by Rogers and used as one of his 'case illustrations' in the third edition of *Diffusion of Innovations* (discussed below) to highlight the influence of networks on the diffusion process (Rogers, 1983). Although it dealt with building products, the study was essentially based on household consumer behaviour, as the solar panels were installed into the roof of existing houses. In this example, it was the visibility of the product that affected its rate of adoption. In a similar way, double-glazing was promoted in Britain by direct selling to householders.

The maintenance and upgrading of buildings, of which the above are just two examples, are operations that encompass both professional activities, where corporate clients are involved, and the ordinary consumer who may be buying direct from the manufacturer or working with the assistance of a builder. These may involve very large volumes of sales and so be significant developments in the industry. In the professional field, flooring and partitioning are the major product types involved in retrofitting. While these kinds of products are innovations, the upgrading of commercial properties is often carried out by a quite different group of professionals from those involved in new buildings, so that this and the domestic consumer market lie outside the scope of this study.

Few studies have considered the behaviour of the designer of buildings, i.e. the specifier of new building products, in any detail. The starting point for historians who are looking at the new ideas of particular individuals has been the act of adoption, attempting to explore the possible origins of each idea. The implicit assumption is that what is of interest is the process of transmission of ideas that are actually used. But one might equally ask why ideas are not adopted. If an idea already exists for a long time before it is more generally adopted, then an understanding of the adoption process also needs to consider why it was rejected by those who were previously aware of it. As Yeomans (1992) has shown, at one time a problem for the adoption of technical innovations was the conservative behaviour of apprentice trained craftsmen who were likely to cling to what they already knew rather than to adopt the unfamiliar. Sometimes change only occurs as a result of external pressures. Yeomans, both his recent study of concrete design (1997: 102–127) and his study of timber specification practice at the turn of the nineteenth century (1989), and Bowley's study of developments in frame structures, showed that professionals have behaved little better. If we are to understand these adoption processes, then our attention must be focused on professional behaviour.

Today, few buildings use only the simple, basic materials, such as bricks and timber. The vast majority use a wide range of manufactured products. Even basic components such as the screws and nails used to fasten other components together have undergone improvements that make them quite different from their predecessors. The principal source of information for the designer on these new products is the extensive body of descriptive material produced by building product manufacturers, i.e. their trade literature, which is still the basic starting point even though electronic forms of information are becoming available. Therefore, the natural starting point for any enquiry is the relationship between this information source and the behaviour of the specifier. The first questions concern how this material is made available to the specifier, i.e. the nature of the communication channels between manufacturers and designers. How readily is material available during the design process itself, and what is the behaviour of the designer in using this information? It is with the availability and use of the material in the design office that we need to begin.

Well managed offices will commonly have a library of product information; a central resource that the members of the office can and are expected to use. However, experience shows that many designers will have their own collection of product information, even though this may be in contravention of office policy. In spite of the availability of this resource, it would be wrong to assume that it is part of a designer's normal behaviour to search the product literature to find just the right product for the particular task that he or she has in mind; instead, behaviour is rather more conservative, relying upon what the designer is already aware of. Moreover, office libraries and personal collections are themselves highly selective.

Because designers tend to use familiar products, it may be some time before they even become aware of more recently developed alternatives. If this conservative behaviour results in the formation of a personal palette, how does a building product that is perceived as new by a specifier working in a designer's office get to be specified in preference to one already familiar? There has been little research into the process by which specifiers become aware of products with which they are unfamiliar or into the way in which these products may be adopted in the decision making process that follows. It is necessary to look outside building to fields where the selection of products has been studied to obtain a better idea of this process. For example, to examine how a product is adopted, a natural starting point would be to look at marketing literature, in particular that concerned with consumer behaviour (e.g. Chisnell, 1995). The problem with this is that it is concerned with 'new' products, i.e. those recently launched onto the market. While specifiers may consider building products that are new to the market, they may also consider products that have been available for some time, but which they have only just become aware of. This can occur when a specifier is faced with a new kind of problem and needs to use a range of products of which he or she has no previous experience, simply because they were not required for previous jobs. Then, even those products that other specifiers have knowledge and experience of might be new to this particular person.

The generation of innovations

For a product to be diffused, it must first be developed, manufactured and launched onto the market. This subject area has been covered extensively, from invention (e.g. Gilfillan, 1935), through product development (e.g. Bradbury, 1989) to marketing (e.g. Midgley, 1977; Druker, 1985), and a new idea or new process adopted by the manufacturing industry has been described as a technological innovation (Utterback, 1994) or a process innovation (Davies, 1979). These innovations are concerned with the introduction of new machinery or production methods and their effect on productivity and, because of this, they tend to be studied by economists such as Bowley (1960). Studies concerned with the manufacture of building materials have been carried out by Davies (1979), who included a study of the brick making industry in his work, while Layton (1972: 80–93) investigated the introduction of the float glass process by Pilkingtons.

Parker (1978) referred to the development of new products as the innovation process, and divided the process into four aspects: invention, entrepreneurship, investment and development. Other authors, such as Bradbury (1989), have made a distinction between the initial idea (invention) and the innovation process, which covers all stages of a product's development up to, and including, its launch onto the market. The generation of innovations ends in a decision by the manufacturer to market the product to potential adopters (Rogers, 2003). It is the decision to market the product that is the start of the diffusion process (Fig. 7.1).

Apart from innovation that has come entirely from manufacturers, there is a long history of architects' involvement with these initial phases of the process. Holden (1998) has shown how designers of Lancashire cotton mills in the nineteenth century were concerned with the development of fire proof flooring, while Saint (1987) describes how, a century later, those involved in the post war school building programme involved themselves directly with manufacturers in the design of flooring, sanitary ware and furniture for their buildings. Indeed, there may be a complex relationship between the development of architectural ideas and that of suitable products through which these ideas may be realized.

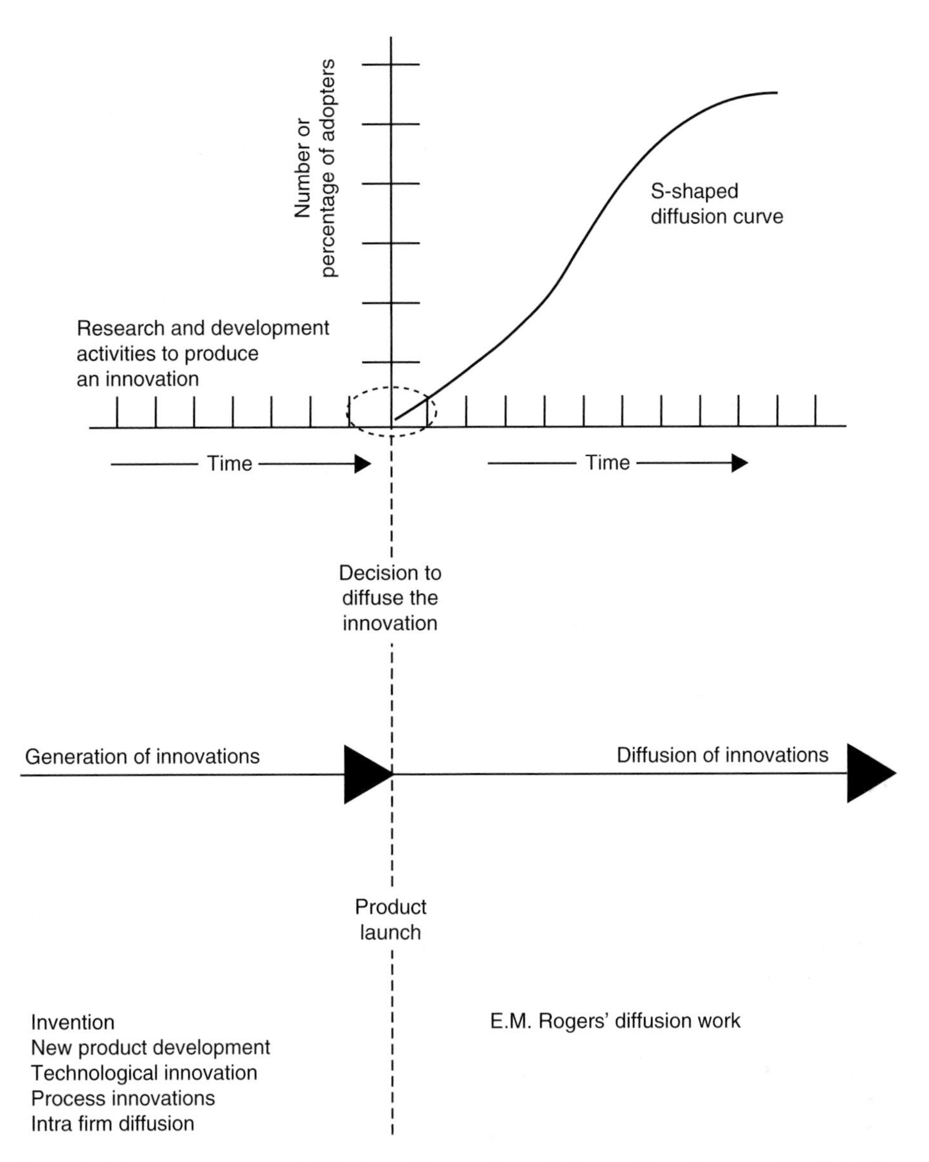

Fig. 7.1 *The generation and diffusion of innovations (adapted from Rogers, 2003: 114)*

The initial ideas in the development of cladding systems were undoubtedly architectural, but they could not have been realized in the way in which they were without the active involvement of manufacturers of cladding components. What is not clear is the extent to which architectural ideas have played a part in the development of other building products. As an example, we may cite the development of devices for fixing brickwork to the structure of buildings as architects in the 1970s began to treat this material in a far more plastic manner. Here, too, was a prima facie example of architectural demand driving the development of building products.

There is a chicken and egg issue here associated with visibility. Products such as those used to achieve suspended, non structural brickwork shapes that derive from a particular architecture are the kind of items that can be produced only as a result of demand pull because they depend on the adoption of a particular architectural fashion. Here, the innovation is the architectural fashion, but one whose wider adoption is dependent on the availability and use of innovative building products, with all the attendant risks both to the specifier and to the manufacturer. The issue of visibility is that while the product may not itself be visible, it has visible effects. If brickwork is used in a sculptural way like concrete, then, to any architect, there is an intimate connection between the visible form and the kind of products that, although invisible, must be used to produce that form.

There is another possible process to consider here, which is the extent to which building designers are able to persuade manufacturers to produce the things that they want. Banham (1969: 204) cited private correspondence from J. R. Davidson, who claimed that in the late 1920s, it was difficult for architects to persuade manufacturers to produce the kind of light fittings that they wanted. More recently, Oostra (1999) has presented a case study of work carried out in the Netherlands to develop window mullions using a new kind of material; new, that is, for window mullions, but widely used in the manufacture of sports goods. In spite of the successful development of these components for a particular project, the idea could not be developed more widely because of lack of interest on the part of the manufacturer. The manufacturer who used the material in the production of sports goods seemed unwilling to enter the building industry, perhaps on the principle of 'cobblers and lasts'. The difficulty is that one is always dependent on such anecdotal evidence, and there are insufficient well researched case studies to know what kinds of conditions favour the uptake of designers' ideas by manufacturers.

The significance of the window mullion example is simply that it shows that there are architects who are not overtly adverse to innovations. Quite the contrary, as architects see themselves as innovative, one might expect them to be willing to embrace the new ideas of others. Nevertheless, Bowley's (1966) studies have given us an image of an industry that is resistant to change. It is a matter of commercial prudence that produces this conservative behaviour. At the same time, one must assume that manufacturers are conscious of the need for their products to be acceptable to architects and therefore to be responsive to their comments. If this is a significant aspect of a manufacturer's marketing methods, then it should only be a small step from the modification of products to take account of feedback from architect to the development of new products to meet their needs. However, the extent of this as an important generator or modifier of new building products has not been studied and is beyond the scope of this investigation.

Definition of 'building product innovation'

So far, the word 'innovation' has been used rather loosely. *The Concise Oxford Dictionary* (1990: 610) describes the verb 'innovate' as 'bring in new methods, ideas, etc.; make changes', and 'innovation' and 'innovator' as 'make new, alter'. The synonyms listed against 'innovation' in *The Oxford Thesaurus* (1991: 223) are 'novelty; invention; modernization; alteration'. The word innovation is used in different ways by different authors to mean different things. Even within a single industry or single profession, innovations can be of many different kinds, concerned with new ideas, new products or new methods. Furthermore, authors concerned with different subject areas, such as economics, politics, sociology, design, engineering, corporate management, marketing and consumer behaviour, all use the word differently. In architectural literature, the word innovation tends to be used to describe either the design approach of the architect or the appearance of the finished building. For example, architectural journalists often refer to the design of the building as 'innovative' or state that the architect has worked in a manner regarded as 'innovative' by his or her peers. In architectural literature, therefore, the word is often used as a substitute for 'creative' and does not have the same meaning as the word in diffusion literature.

Bowley divided innovations into two main groups, 'those that change the product and those that affect costs and availabilities' (Bowley, 1960: 25), and was concerned with innovation as viewed by the consumer, the building user. She classified innovations in a range, from those that result in new products (not substitutes for existing products) to those that lead to products that are, from the viewpoint of the consumer, no different from existing products (a perfect substitute). Bowley went to great lengths to classify innovations (pages 25–43 of her study), concluding with the observation that '. . . there are innumerable ways of working out classifications of innovations, and the advantage of one rather than another depends on the particular purposes of the study' (Bowley, 1960: 43). Others, such as Slaughter (2000), have used different terminology to identify five types of construction innovations (incremental, architectural, modular, system and radical), which helps to highlight the need for clear definitions. To aid clarity, the terminology used here follows the tradition of the large body of diffusion of innovations literature.

It is the body of work on diffusion of innovations (Rogers, 2003) that is concerned with an individual's reaction to new ideas and which examines the mechanisms of adoption or rejection. This work treats an innovation as something that is perceived as new, whether or not it is in fact new. It is the newness of the idea to the recipient, rather than the length of time that it has been on the market, that sets diffusion literature apart from marketing literature. Because diffusion theory is concerned with the factors that influence the rate of adoption of ideas or products that are perceived as new by the receiver of the information, the potential adopter (in this case, a specifier working in a design office), it provides a general model that is more relevant to the behaviour of the designer and specifier.

Rogers is concerned with the total population of a social system, i.e. all potential adopters, and has defined innovation as a product or idea that is new to the recipient, regardless of how long it has been available. Since the research reported here is concerned with the perception of specifiers working in design offices, for the purpose of this work, the definition used by Rogers (2003) can be rewritten as:

> An innovation is a building product that is perceived as new by a specifier. Whether or not the product has been recently launched onto the market is not important, it is the perceived newness of the product by the specifier that determines his or her initial reaction to it.

Diffusion of innovations research

If the economic process depends on the capacity of industry continuously to develop new products and new processes (Druker, 1985), it follows that it is equally dependent on their adoption by the consumer, and it is this process that has been explored in diffusion research. The spread of new ideas, practices and products within a social system is known as the diffusion of innovations. In simple terms, diffusion studies are concerned with the communication of an innovation to a social system over time, described by Rogers (2003: 5) as:

> . . . the process in which an innovation is communicated through certain channels over time among the members of a social system. It is a special type of communication, in that the messages are concerned with new ideas.

The history of diffusion research is well documented (Rogers, 1995, 2003), the main subject areas covered being anthropology, early sociology, rural sociology, education, public health, medical sociology, communication, marketing, geography and general sociology. Published in their own field, these diffusion studies were concerned with the uptake of both innovative ideas and innovative products.

Tarde (1903) was recognized as the first to investigate the adoption or rejection of innovations. His publication *The Laws of Imitation* identified several of the main issues of diffusion, from the S shaped curve to the important role of the opinion leader in a social system. While the influence of Tarde is still present in diffusion studies, the start of diffusion research dates back to the 1930s, with the majority of early work undertaken by rural sociologists. Given this tradition, it is not surprising that a rural sociologist was the first to publish a comprehensive book on the subject with *Diffusion of Innovations* (Rogers, 1962), since which time, Rogers has continued to publish in the field.

The first edition of *Diffusion of Innovations*, which summarizes diffusion literature, has been described as a 'benchmark study' by fellow researchers (e.g. Brown, 1981). It consolidated the work of 405 separate publications, including twenty-seven of Rogers' own, provided generalizations and definitions that could be used universally, and provided a single model of the diffusion process that others could, and did, use. Rogers continued to update and revise his work as the number of individual diffusion studies increased and a second edition, *Communication of Innovations: A Cross Cultural Approach*, co authored with Shoemaker (Rogers and Shoemaker, 1971) had a strong emphasis on the communication process, reflected in both the revised title and the influence of Shoemaker. The third edition marked a return to the single author and original title. By the time the fifth edition of *Diffusion of Innovations* was published in 2003, there were over 5200 independent publications, of which approximately 75 per cent were empirical studies.

The main principles of diffusion work categorized by Rogers have been used by the majority of subsequent diffusion researchers, whose work has ranged from simple models to sophisticated models based on complex mathematical formulae. For example, in geography, Brown (1981) uses the work of Hagerstrand (1969) and Rogers to develop his own paradigm of diffusion across the landscape: spatial diffusion. Others have either applied the model to different fields, such as manufacturing and product development, or concentrated on a small part of the process. For example, Foxall (1994) concentrated on the characteristics of adopters, Gatignon and Robertson (1991) concentrated on inter personal communication in the development of their consumer diffusion paradigm, and Valente (1995) worked with network diffusion models.

Rogers has shown that the innovation will have a number of perceived characteristics, which will influence its rate of adoption, but also that the individuals in the social system who are exposed to knowledge about the innovation will themselves have different characteristics. Based on empirical findings, the individuals in a social system have been classified into five categories, ranging from the first to adopt, the innovators, to the last, the laggards, according to their degree of innovativeness (Rogers, 2003). The process through which the adopter passes, from first exposure to the innovation to a decision as to whether or not to use it, is known as the innovation decision process, and this is discussed in more detail below.

The diffusion of trussed rafter roofs

The transmission of trussed rafter roof technology from the USA to Britain has already described in some detail (Yeomans, 1998b), but will repay an abbreviated retelling in this context. Trussed rafter roofs depend on toothed plate connectors, an invention for joining timbers together that comprises metal plates with teeth pressed out of the plate. The plates are then pressed onto timbers so that they overlap two or more pieces at a joint, the teeth biting into the timbers and thus joining them together. The plates are, in effect, the equivalent of gusset plates that had long been used in steel roof trusses. They allowed several timbers in the same plane to be joined together, and the ability to do this meant that simple trusses could be formed using very small-section timber. The result was a return to the common rafter roof, i.e. a roof without purlins. Moreover, as this was a factory operation it removed work from the building site and so had the additional advantage of speeding up building operations.

The money for the patentees of the plates was in their sales and in the licensing of the computer based methods used to design the trusses. The marketing strategy was to license timber merchants to use the design software and to sell trusses made using the plates. This made the builder's job simple because he could specify the roof span, pitch and loading and the timber merchant would do the rest, delivering completed trusses to site. Having achieved success in the USA, the plate manufacturers looked to the market in Britain, where there was also a rapid uptake of the system. One can see that this involved some restructuring of the industry as timber merchants became designers and manufacturers of a building component, while the need for carpentry skills on site was reduced, some might say eliminated. However, after a few years of the use of these trusses in England serious problems arose when it was realized that a large number of roofs were showing signs of distress. There was also a number of collapses of gable walls that could be attributed to poor construction using these trusses, but the real wake up call came when a major roof collapsed during construction.

The problem was that there was a fundamental difference between English and American roofs that had not been recognized. Roofs in the USA use plywood sheathing nailed to the rafters to support shingles, an asphalt-based roof covering. The plywood thus ensured stability of the roof. In England tiling battens failed to ensure stability in the same way because they were inadequately attached to the gable wall. Moreover, without the purlins, the gable wall was inadequately restrained against wind load, which had accounted for some of the failures. Scotland seems to have avoided the general problem of instability of the roofs because of the common practice there of using sarking boards.

There were other causes of distress that were particular to individual building designs, but they could generally be attributed to a failure to consider adequately the structure of the roof and its relation to the rest of the building. Here was a technology perfectly transmitted but only in part, and as a result misunderstood and misused. The problem has now been rectified, partly as a result of changes to the design code. However, it is worth noting that during the period when there was growing concern about these failures an informal test was carried out at a workshop held to discuss the problem of the knowledge of those attending. The respondents were mostly architects and clerks of works, who one might expect to be familiar with the available literature. The results showed widespread ignorance of some of the fundamental requirements of the Code of Practice current at the time. The technology was perceived as being simpler that it in fact was.

The issues have been simplified here somewhat because, to some extent, this was a problem waiting to happen. Other changes in the construction of houses in Britain had reduced what had been a fairly robust structure to little more than a thin masonry shell, relying on the floor and roof for its stability. Moreover, poor on site storage of trusses often resulted in their distortion, which exacerbated their tendency to fail.

Rogers' analysis of innovation diffusion processes describes the trussed rafter roof episode fairly well. His change agents in this case were the marketing organizations of the truss plate manufacturers. The failure was not simply that adopters did not appreciate the technical complexity of the roofing system; it was also the failure of the change agents to take full account of the compatibility of the system with indigenous building practices. This is a recognized problem and Rogers describes instances of its occurrence. The trussed rafter roof as used in America was incompatible with English construction practice because the latter does not use plywood sheathing.

This roof problem was first addressed by the plate manufacturers by issuing leaflets with all deliveries of trusses that showed builders the correct method of both storing and handling the trusses and erecting the roof. These leaflets also showed the additional timbers needed to ensure its stability. The truss plate manufacturers were clearly anxious to avoid discontinuance of the innovation because if that happened an older alternative was waiting in the wings to make a comeback. This was the Timber Development Association (TDA) roof, developed after the Second World War to eliminate the need for heavy purlins and promoted through a set of cheaply available standard design sheets. Builders had found these reliable and might well have returned to their use. While the leaflet initiative was designed to advise the unsophisticated house builder, other action was needed to educate the professionals. They had similarly failed to understand just how the roof was working and its interaction with the wall, but also often failed to grasp the level of information that the manufacturer needed to be able to produce a sound design. Here we are outside the scope of Rogers' work because most studies have looked at the obstacles to change, whereas in this case there was the overenthusiastic adoption of an innovation that was incompatible with both current practice and apparently the level of education among adopters.

The innovation decision process

At the heart of all diffusion research is the adoption process, in which an individual or a group of individuals becomes aware of an innovation and reacts to it, by either adopting or rejecting it. Reaction to the innovation is not an instantaneous act, but a process that continues over a period of time with five stages known as the innovation decision process (Rogers, 2003) (Fig. 7.2). It is the cumulative effect of adoption of an innovation over time that results in the classic diffusion curve. The innovation decision process has parallels with the specification process and provides a useful theoretical model to explore how specifiers react to building product innovations. The model is discussed briefly from the context of a specifier working in a professional office.

The specifier will pass from first exposure to information about the innovation (knowledge), through a period of gathering more information to consider its characteristics (persuasion), to making a decision to use or reject the innovation (decision), to construction on site (implementation) and the intention to use the product again (confirmation). The innovation decision making process is important because the individual selects a *new* product over one previously in existence. Thus, the *newness* of the alternative is an important aspect of the innovation decision making process.

This process is going on when the specifier is detailing the building; as such, it will be influenced both by the particular design project being worked on and by the amount of time available to the specifier: it is not carried out in isolation but as part of the detail design decision making process. As noted earlier, the design of a building will involve a matrix of decision making that will vary in complexity as the design progresses. A number of specific decisions is taken at each stage that in turn influence or determine those that succeed them. Although the Royal Institute of British Architects (RIBA) Plan of Work indicates a theoretical framework for decision making, in practice the sequence of decision making is not always adhered to. Some stages might be skipped and there is usually constant reassessment and iteration of the design problem. Each specifier will have his or her own subjective perception of the problem, based on past experience and possibly the past experience of the office (i.e. previous jobs of a similar nature). Specifiers will attempt to find the action that will be satisfactory and not necessarily optimal in meeting their objectives. As discussed above, the decision making will be influenced by the individual's personal characteristics (status, age, experience, personal values, etc.), the situation (such as the type of building or the stage of the project) and the amount of time available in which to complete the process (degree of urgency). The specifier will also be working within a set of parameters set out in the briefing documents and early conceptual design. Thus, the innovativeness of the specifier may be influenced by project characteristics and hence may vary between projects. Furthermore,

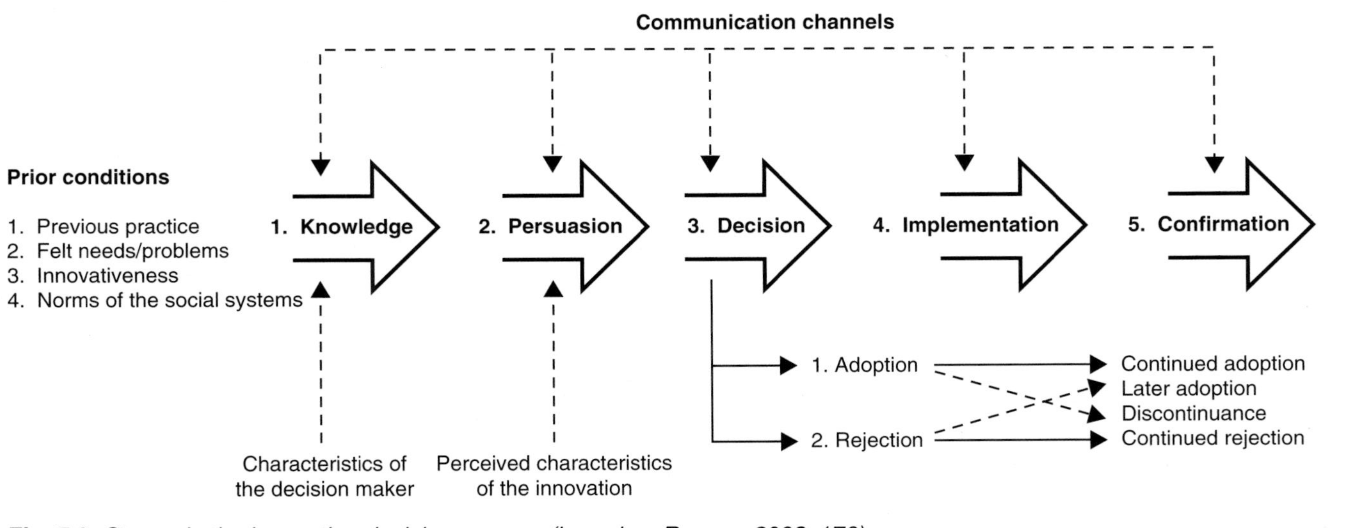

Fig. 7.2 *Stages in the innovation decision process (based on Rogers, 2003: 170)*

the contribution from people outside the immediate social system cannot be ignored.

Before this process starts, however, the specifier must have identified a problem that cannot be resolved from the information contained in the collection of favourite products, hence triggering a search for information and the start of the innovation decision process. The challenge was to try to observe this process.

Research methodology

Despite the extensive body of diffusion work there are very few empirical diffusion studies within building. The extensive bibliographies contained in the five editions of Rogers' work do not contain any specific references to studies in this sector, other than a couple of studies concerned with the manufacturing of building products (e.g. Davies, 1979). A similar observation can be made of Musmann and Kennedy's (1989) survey of the literature. Here, only one study (Leefers, 1981) addressed construction and this was concerned with the economic factors that influenced the diffusion of both wood particle board and southern pine plywood. These bibliographies help to illustrate the dearth of diffusion research in construction (Emmitt, 1997), exceptions being work by Larsson (1992), Koebel et al. (2004) and Larsen (2005).

Larsson (1992) examined the adoption of new building technologies on Swedish construction sites. This helped to emphasize the different focus between manufacturers (product focus) and contractors (project focus), which resulted in communication difficulties between the parties. Larsson used the Rogers diffusion model to understand better the decisions made on the construction site by site managers, collecting data from ongoing projects through interviews with the project participants. In this study the contractors were making decisions about purchasing products from suppliers and manufacturers. These decisions were taken with regard to the profitability of the project and, according to Larsson, the adoption of new products was highly dependent on the site manager. Larsson also found that the site manager was struggling to find adequate time to make decisions.

A study into the diffusion of innovations in the US residential house building sector helps to illustrate some of the characteristics particular to construction (Koebel et al., 2004). Data collection was by means of a questionnaire survey of house builders to try to understand why some appeared to be more innovative than others. Although Rogers' diffusion terminology was used in the work the authors did not attempt to apply their data to the Rogers model, nor did they look at the specification process.

Larsen (2005) used interview data to develop a hypothetical diffusion journey for actors working in the UK construction sector. The model developed by Larsen has similarities to Rogers' innovation decision process, although the importance of personal awareness thresholds is better articulated.

Given that there has been virtually no diffusion work on the building sector, it may be helpful to look for parallels in other areas; i.e. situations that may have some similarities with building. There are two distinct areas within diffusion literature that might be considered to have some relevance: research into the diffusion of new products (e.g. Bass, 1969; Mahajan and Wind, 1986) and research into the adoption of drugs by medical doctors (e.g. Coleman, 1966). Although work on the diffusion of new products has drawn on the Rogers model, it is specifically concerned with products that have been launched onto the market recently; and given the special nature of design work (discussed above), it is unfortunately not relevant to this work. Research into the adoption of drugs by medical doctors is more relevant because it examined the adoption behaviour of a professional group that has parallels with building designers. Doctors have a duty of care when selecting drugs on behalf of their patients, just as architects and engineers have a duty of care towards their clients when selecting building products. Unfortunately, the prescription of medical drugs is a relatively simple affair compared with the specification of building products, as demonstrated in Chapters 8–10.

Rogers has criticized diffusion research for the manner in which data have been collected in the majority of diffusion studies. The research has been based on retrospective data collection, i.e. an innovation that has been adopted and diffused is then traced back to examine the reason for its adoption; usually by interviewing the adopters some time after their decision to adopt the innovation was made. This method of data collection suffers from the disadvantage that it focuses only on successful innovations. It is difficult to conduct research into innovations that failed to diffuse because there is less information. One way of overcoming this is to collect data while the adoption process or diffusion process is underway, allowing the study of unsuccessful innovations. Rogers (2003) has suggested that researchers need to acknowledge that rejection, discontinuance and reinvention occur frequently during the diffusion process; that the broader context in which the innovation may diffuse should be investigated; and that the motivation for adoption needs to be addressed. This emphasizes the need to collect data from live design and construction projects.

Previous research into the selection of building products has tended to rely on asking specifiers what they did (Barbour Index, 1993, 2000, 2004, 2006; Chick and Micklethwaite, 2004; Koebel et al., 2004) and/or asking them to record their decisions in research diaries supported by interviews (Mackinder, 1980). This body of work is based on how specifiers claim to act, and while such approaches are valuable they are prone to difficulties. This is, first, because professionals tend to portray themselves as they wish to be seen (Ellis and Cuff, 1989), so that their account of how they act may not necessarily reflect what they actually do. Secondly, they have a rather poor ability to remember and recount all the steps in the process (Yeomans, 1982). Thirdly, asking specifiers to record their own behaviour in diaries naturally raises their awareness to specification decisions and might influence their behaviour. Thus, asking busy specifiers to recall their actions may not be the most accurate or appropriate method of data collection when trying to understand the actions behind specification decisions.

In looking at building products, evidence of what actually happens during the specification process was required, because it is the innovation decision making process that is important. A detailed analysis of drawings and schedules would provide a list of products actually specified, while variation orders issued during the contract would confirm any changes to specified products that have occurred during the contract. However, this approach would not provide any evidence of how and when the specifier became aware of the building product, nor would it provide any insight into a specifier's innovation decision process. That is because there is no reason to record this in the normal course of work. Furthermore, it would be both time consuming and extremely difficult to identify which products were new to the specifier without resorting to an interview.

Method

Since specifiers must be aware of a building product in order to specify it, communication of information about building product innovations to the potential specifier is critical to their adoption. An enquiry was therefore conducted into the communication of such innovations to the architect's office with attempts to measure the specifiers' reaction to these products, identifying the factors that led to either their adoption or their rejection. This is an aspect of the design process that has been largely overlooked, or ignored, by the design methods authors, i.e. the detail design process during which building products are specified. The intention of the enquiry was to use the Rogers diffusion model as a structure for the work, the objective being to try to establish how specifiers respond to building products that are new to them. It did not attempt to address the level of innovativeness of architects, nor was it directly concerned with how the diffusion of innovations affects the design, function or appearance of the building. It was an attempt to understand the behaviour of architects, or other specifiers (acting on behalf of their clients), towards building products that were new to them, i.e. it was primarily concerned with the innovation decision process.

The first stage was to conduct a questionnaire survey of specifiers to collect some background information and to see whether they had changed their views since Mackinder's work was published. The results supported Mackinder's work and helped to emphasize the need for ethnographic work. Data were collected from within architectural offices using direct non intrusive participant observation, to study an organization in its natural state, i.e. undisturbed by the researcher (Hammersley and Atkinson, 1995). This was supplemented with semi structured interviews and analysis of documentation produced by the designers in the course of their work. (Drawings, written specifications and file/diary notes were used to check the validity of the observations and comments recorded in the interviews.) The research adopted what Gold (1969) describes as the 'complete participant' approach, engaged fully in the activities of the organization under investigation. The goal was to interpret the behaviours of the social system being studied (e.g. Rosen, 1991; Nason and Golding, 1998), an approach adopted successfully in earlier research based in architectural offices (e.g. Cuff, 1991). Although findings are specific to a particular context at a particular point in time, they help to illustrate the specification process in more detail than other research techniques allow. In particular, the pressures placed on specifiers, such as the lack of time to complete the task and pressure from other members of the design process to influence the decision making process, have become more apparent.

8 Becoming aware of new products

Specifiers may become aware of new products via a variety of routes: through the marketing activities of manufacturers, through information passed to them by colleagues in their office or by attempts by those outside the immediate design team to influence their choice of components. But those that are of most concern to us here are the methods used by manufacturers in marketing their products. Clearly, it is important to both manufacturers and specifiers that these channels be used effectively; manufacturers want to sell their products and specifiers want the information to be at hand when it is needed in design. Thus, the effectiveness of these channels of communication needs to be considered from the point of view of both. Manufacturers may wish that specifiers behaved in ways that would make better use of their marketing material than they do at present, but this is unlikely. Their behaviour is constrained by the circumstances of their design tasks, by the management of the offices in which they work and by the limits of their education. In this relationship they are the customers whose needs have to be met and it is the task of the manufacturers to adapt their marketing strategies to these needs and behaviour patterns. The purpose of this chapter is to consider how that might best be done.

Early clues

One of the authors attended a focus group, or group discussion, arranged by a commercial market research firm. Their aim was to investigate the reading habits and behaviour of architects with a view to restructuring the editorial content of a long established weekly publication, which had a strong practical bias in its content. It regularly reviews building products and reports on the technical aspects of the buildings that it describes, as well as carrying advertising space for manufacturers. The nine architects who took part in the group discussion comprised two partners/directors in medium sized to large architectural practices, five associates in small architectural practices, and two solo practitioners. All of the offices were located in the same geographical area, but none of the architects knew one another personally. Although a very small sample, this suggested that architects' offices are not linked by a network, so that the possibility that knowledge of building product innovations might be diffused by social interaction between design offices is unlikely.

Apart from the two solo practitioners (who carried out all aspects of the architect's duties), the architects said that they were not involved in building product selection themselves, but were responsible for supervising or overseeing the work of less senior members of their office who were. Product specification was seen by them as the least glamorous and most tedious job in the office, as well as being a particularly time consuming task, and tended to be delegated to the lower paid members of the office, thus confirming Mackinder's observations.

When questioned on the subject of specification, all stated that the office selected familiar products. This was reinforced by the use of master specifications and standard details that were applied to all jobs and altered if necessary to suit the specific criteria of a particular project. These specifications and details had been developed over a number of years and contained what each practice considered to be the best materials or products for their purpose, based on previous experience. Products included in master specifications had been specified and built into previous projects and so were both familiar to the office and known to perform the function for which they were specified. The entire sample said that they preferred to specify products by proprietary name rather than by using performance specifications and were very reluctant to change products once specified. This was because of the time required to assess the characteristics of any new product.

All said that a product that was new to the office would only be specified if it was supported by guarantees, such as British Board of Agrément certificates and/or conformity to British Standards. Manufacturers' guarantees were seen to be essential in reducing (perceived) risk, thus minimizing the possibility of an insurance claim for specifying a defective or unsuitable product. Furthermore, they said that they would only use new building products if absolutely necessary, preferring to wait until someone else had used it. Products that had been specified on past building projects and had not performed as stated, or had failed, were deleted from the standard specifications and effectively blacklisted.

The participants said that products could not be compared on a cost basis because the price of a product was rarely known at specification stage. This was because manufacturers were reluctant to disclose prices of products until after they had been specified. Feedback on overall project costs was provided by the quantity surveyor (QS) and, sometimes, the contractor when he wanted to change a specified product for a cheaper alternative. This allowed the architect to build up an elementary knowledge of costs. The two solo practitioners, who carried out work for small scale developers and builders, claimed to have a better knowledge of costs because they had to work without the services of a QS. They claimed that they rarely used products that were new to them because of the time required to investigate them fully and the threat of legal action should they fail.

The primary source of awareness of new products was the office library, which, for larger offices, was administered on a part time basis by a servicing agency, with varying degrees of input from other members of the office. This helped to ensure that the library was updated, usually on a monthly basis, with manufacturers' up to date literature. After the library, the trade journals were the next source of reference. The journals were generally used when looking for sources of inspiration for the design of a particular building type, with less emphasis on product selection. They all said that they rarely took any notice of advertisements in the journals, but did say that they would occasionally send off for more information using a 'reader reply card'. Trade fairs and exhibitions were rarely attended because they were seen as a waste of time.

The architects were then asked about the sources that they used to select products. Past experience of the architect and of the specifiers in the office was the main source of information for the entire sample. Six said that they occasionally used trade literature in the library, while one sometimes followed up articles on projects featured in the journals. In a general discussion, prompted by the organizer, the sample said that occasionally, they would use a product recommended by a client or a consultant. Trade representatives would only be invited into the office once a genuine interest in the product had been established and only where it was relevant to a current job. If the representative called speculatively, they were usually asked to leave their literature, but rarely seen by specifiers in the office because of the lack of time available. In general, the architects were dismissive of the trade representative.

The focus group discussion was instrumental in identifying a need for some background information on how specifiers believed they were behaving. It was evident at an early stage that there was little published material that related directly to this subject, and the simplest way to begin was to gather some preliminary information using a postal questionnaire (see Appendix). The questionnaires were addressed to senior architects and partners of the architectural practices.

The purpose of the postal questionnaire was to test ideas that were being developed in the theoretical model. A traditional form of contractual arrangement was assumed in this, and that specifiers selected building products by brand name rather than by generic terms. Therefore, it was important to establish the extent to which these assumptions were borne out in practice. Answers to three questions supported a general model based both on the use of traditional contracts (awarded on the basis of competitive tendering) and on product specification by brand name. Rogers' (2003) innovation decision process is a useful model in helping to understand how specifiers become aware of new products (see Fig. 7.2) and the results of the questionnaire survey are discussed within this framework. The questionnaire survey was also supported by a small number of semi structured interviews with manufacturers and specifiers, and the findings of these are also summarized below. The focus was mainly on the earlier part of the adoption process, since the later stages of implementation and confirmation could only be addressed through observational research (which is reported in Chapters 9 and 10).

Prior conditions

Under the general term 'prior conditions', Rogers identifies four factors affecting the adoption of an innovation: their previous practice, their needs or problems, the innovativeness of the potential adopter and the norms of the social system in which they are operating. These were explored from the specifier's perspective using a questionnaire survey.

Previous practice

Previous practice is associated with the types of contractual arrangements used, preferences for a particular method of product specification and the influence of office policy. Traditional contracts had been used more than any other contractual arrangement; 92 per cent of respondents indicated that it was their first choice, with only 8 per cent liking design and build. It was assumed that the type of contract would influence the manner in which building products were selected, but when respondents were asked directly if this was so slightly less than one third answered 'no', and a similar number 'generally not'. Of course, given the preponderance of traditional contracts, they may not have had sufficient experience of other types to know whether selection of products would have been different or not.

There was a strong preference for the use of precise trade names (brand names), with 85 per cent always or often selecting materials by this method. However, 58 per cent also claimed that they most frequently selected materials by a generic description. The inference is that both methods are used concurrently for different product types.

The presence of an approved list of products was noted in one third of the practices, while about the same number operated a list of prohibited products. Lists of prohibited materials and products were not widely reported, but it was noticed during the group discussion (reported above) that architects were reluctant to admit to the existence of such lists, so that this response should be treated with caution.

Felt needs/problems

The design of buildings is rarely a standard procedure and it is likely that many projects will involve specifiers in a search for information about products that are new to them, simply because they face an unfamiliar problem which cannot be resolved by applying tried and tested solutions. This triggers a search for information and one would presume greater attention to various manufacturers' marketing campaigns. Specifiers usually operate their own gatekeeping mechanism, closing the gate to information when engaged in tasks other than product selection, a passive period, and opening the gate if their palette of favourite products is unable to solve a particular problem, an active period during which information about building products is sought. This is clearly demonstrated in the case studies reported in Chapters 9 and 10. The manner in which the specifier becomes aware of information about a building product innovation is important because it is the first step in the innovation decision process.

In 1981, the *Architects' Journal* summarized the findings of a telephone survey, conducted by a market research company (Walton Markham Associates Ltd, 1981: 380) that asked 100 architects how they obtained information about building products. Their comments were that:

- They obtain product information from journals and by sending off reader enquiry cards for selected products, i.e. initial awareness through the paper literature (advertisements and technical articles).
- Direct mail 'bombardment' was seen as 'a nuisance to be borne'.
- They regarded promotional events and trade fairs as a waste of time.
- They complained that technical literature lacked detail.
- Unsolicited visits to the office by trade representatives were not welcome; the architects said that they were not interested in being sold to; instead, they would go in search of a product when needed.
- All had a central (office) library of building product information, while 60 per cent also kept a personal file of information (palette of products).

The sample size was small, but the finding that 60 per cent kept a palette of favourite products supports Mackinder's (1980) earlier work. Perhaps the most important point was that designers search for information about a product when the need arises. Therefore, awareness is not simply passively acquired: it is actively sought at certain times.

If a specifier does not identify a need to change a particular product from that previously adopted (from the palette of favourite products), his or her personal gate will be closed to trade information, and thus, awareness of building-product innovations is unlikely. However, a specifier who is dissatisfied with a previously used product will feel a need to change, actively seek an alternative, and thus be more responsive to information about new building products. When a specifier wants a specific piece of information, he or she has two alternatives: to consult fellow specifiers in the office (their previous experience and the records generated by the past work of the office), or to search for it among manufacturers' information. Work by Mackinder and Marvin (1982) found that architects choose the latter 'only as a last resort'.

Innovativeness

Allinson (1993) and Gutman (1988) categorized architectural practices by the type of service they provide, dividing them into 'strong idea firms', 'strong service firms' and 'strong delivery firms'. From this, it is reasonable to assume that the type of architectural practice may influence their attitude to the selection of building products. Symes et al. (1995) also reported that 80 per cent of the firms surveyed specialized in at least one building type, such as housing or commercial and industrial buildings. Clearly, such specialization will influence the range of building products that are selected and presumably the skill with which they are able to specify them. Offices will have more experience of products that have been developed for the building types in which they specialize. This makes it particularly difficult to classify design offices within the range of innovativeness identified by Rogers. Similarly, the innovativeness of the specifier is difficult to determine unless one focuses on a single product and surveys a very large sample (which is outside the scope of this work).

Norms of the social system

Norms of the social system relate to the norms of the specifier's office, the temporary norms of the project team and the norms of the construction sector. The norms of the professional office tend to be relatively stable, reinforced by the behaviour of the owners of the business and the use of standard operating procedures. The project team is much less stable, and the norms can change relatively quickly as new members enter and leave the project team (Emmitt and Gorse, 2007). This can have an influence on the behaviour of the specifier. It should be noted that the specifier is usually working to project milestones for completing work and so time pressures may have a bearing on their behaviour.

The norms of the social system appear to be related to the use of favourite products and manufacturers, as identified by Mackinder, Walton Markham and some of the Barbour Index reports.

The knowledge stage

Obviously, the specifier must be aware of the building product innovation in order to consider it for adoption. Rogers has accepted that this is a chicken and egg problem – which comes first: the need to change or awareness of the innovation? In the diffusion of a medical drug (Coleman, 1966) the doctors did not seek information about the drug, but gained knowledge of it from advertising and salespeople, thus acting in a passive manner. Only when they were aware of the drug did they seek additional information about it. While in some other respects the prescription of drugs has been likened to the specification of building products, the behaviour of doctors and that of specifiers is quite different in this regard. While doctors appear to be making themselves aware of developments in drugs, specifiers are not similarly keeping up with developments in building products. There is a simple explanation for this apparent unprofessional approach; it is simply that there are too many new products for the building specifier to cope with. Bullivant (1959) noted that architects' need for information is directly related to the stage, or timetable, of their projects. Then they may be responsive to information about products, but at other times information is likely to be ignored, i.e. specifiers will operate their own gates (Fig. 8.1).

Rogers (2003) has argued that individuals will seldom expose themselves to messages about an innovation unless they feel a need for it. Furthermore, even if these individuals are exposed to such innovation messages, there will be little effect unless they perceive the innovation as relevant to their current needs and consistent with their existing attitudes and beliefs. This is known as selective exposure (and selective perception; Hassinger, 1959) and implies that the need for an innovation will usually precede awareness of it. For example, if the designer is faced with an unfamiliar detailing problem that cannot be solved through reliance on familiar solutions and products then there is a need for an innovation. According to Rogers, awareness is related to individual level of attentiveness at the time of exposure to the innovation. Presumably, the specifier's level of attentiveness is higher when actively engaged in detailing the design and writing the specification, than in some other stage of the design process, as previously discussed.

There is an important issue for manufacturers here. The facts as presented are that manufacturers constantly bombard specifiers with information, but the gatekeeping research (Emmitt, 2001) demonstrated that this is an extremely inefficient way of operating because, even if the specifiers were prepared to consider this kind of product, the information may not reach them. Building product manufacturers who simply bombard the design office with direct mail can only hope that the specifier will become aware of the innovation by accident (due to random or non purposive activities), or that they will be lucky, and the information will reach the specifier when he or she is actively searching for an alternative product. Should this fail, they hope that the direct mail literature will be filed in the office library (or the specifier's personal library) for possible reference at a future date. Although trade literature may have been in the office library for some time (and may be out of date), it is still a potential source of awareness.

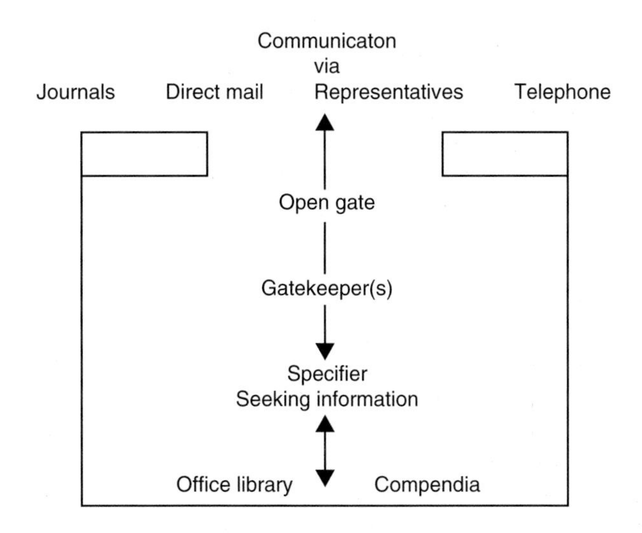

The 'active' specifier

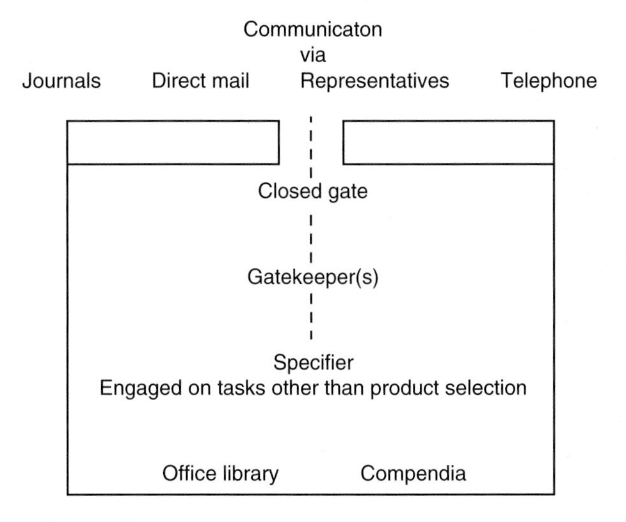

The 'passive' specifier

Fig. 8.1 *Active and passive awareness period*

Reaching the architect during the passive period is not going to be achieved through mailing, and the trade representative is not going to get through the door in the normal course of events, which leaves only continuing professional development (CPD) as a vehicle. Often, CPD events are sponsored by manufacturers who hope to promote their products, although the ostensible reason for these events is to deal with some technical issue associated with the product type. In such an event, potential specifiers are brought to a state of receptiveness by concentrating their minds on an issue that may not be directly related to the particular product. A group of products and associated issues might be discussed and while the latter might be concerned directly with design, they might equally be concerned with something quite different, such as construction safety.

As an example of this, lead paints have not been used for a long time now because they are poisonous. However, many other products, especially those with volatile solvents, may be toxic in application, and concern for health and safety issues has focused the attention of manufacturers on developing new products that are not toxic. The way in which these formulations have been developed and the implications of this will not be apparent to the specifier, nor perhaps of special interest to him, and yet, it is in manufacturers' interests to make the specifier aware of the improvements as a means of encouraging the specification of their products. This is an aspect of trade information that can perhaps best be conveyed during CPD events when, away from the pressures of immediate design and production problems, and perhaps away from the office itself, the specifier will be able to give his mind to such an issue. The manufacturer's hope here is that the specifier remembers the product when next engaged in detailing and specifying activities, and considers it alongside the familiar products.

The marketing managers of three manufacturers were interviewed and asked how they raised the awareness of potential specifiers to their products. All three had the same strategy, their marketing campaigns relying on advertisements in the trade journals and direct mailing, supported by their trade representatives who introduced the product (with samples and trade literature) when visiting specifiers' offices. Trade literature was also sent out when a specifier responded to an advertisement in the trade journals or contacted the manufacturer directly via their homepages, which was also followed up by a visit from the trade representatives. These interviews were carried out with the marketing managers of large, successful companies who had the resources available to employ a wide range of marketing techniques. Many other, smaller, companies do not employ trade representatives, who are an expensive resource, and so have to rely almost entirely on paper and digital information to bring about awareness among potential customers. Manufacturers clearly need a strategy that links their journal advertising, their technical literature and the use (or not) of trade representatives, discussed below.

What specifiers want and how they behave

The question to address is the extent to which a manufacturer's marketing strategy actually matches the behaviour of specifiers: the notice that they might take of advertisements, the use they make of technical literature and their relationship with trade representatives. The majority of the respondents said that trade journals influenced their design decisions and their selection of materials, although only 35 per cent thought that they were likely to select a material or product on the strength of an advertisement or a technical article in a journal. However, unprompted comments such as 'No, but instigate checking on it' and 'No, further research needed and Agrément certificate to be examined' reinforced comments from the group discussion that awareness was raised by articles and advertisements contained within the journals.

The Internet has brought about changes in the way in which specifiers access trade literature through mass media channels. The vast majority of specifiers' offices have access to the Internet and the Barbour Report 2006 reported that a growing number of specifiers are using the Internet as their first choice when searching for trade information. It found that specifiers, contractors and clients obtained information about new products from various sources:

Trade journals	54%
Internet and manufacturers' websites	37%
Direct from manufacturers	34%
Exhibitions	7%
Directories	3%
Manufacturers' CDs	1%
Colleagues/other team members	13%
Seminars/lectures	5%
Other sources (not identified)	16%

Specifiers also searched the trade literature held in the office library (Barbour Index, 2006), which is still retained by approximately two thirds of the specifiers surveyed.

A high percentage of the respondents (92 per cent) claimed that they consulted trade literature on a regular basis, but more significantly, 80 per cent confirmed that they kept their own file of product information. This is important, because the presence of a palette of favoured products could be a barrier to the awareness and subsequent adoption of product innovations, especially those marketed by a company not included within this personal library. When asked about trade literature, 63 per cent of architects expected to be able to make a full and detailed specification on the strength of product information. Unprompted comments reiterated their expectations, from 'Yes, vital', suggesting that if it was not possible they might look elsewhere, to 'Expect yes, more often though it cannot be done'.

Respondents here were interpreting the word 'expect' in terms of what they want from the material rather than their common experience, i.e. reflecting a pattern of behaviour rather than an expectation about the true nature of trade literature. What specifiers seek is the ability to carry out this part of their job as simply and quickly as possible. They will be aware of the additional time needed to seek out information on the product from other sources, and so the quality of the product information is part of the selection process. What they actually 'expect' in another sense of the word may be quite different. The comments on the actual quality of trade literature suggest that in that sense, their expectations based on experience are quite different.

The majority said that they call a trade representative to assist with the specification: 57 per cent sometimes and 25 per cent always. This underlined the importance of the trade representative as an agent of reinforcement. Again, two unprompted comments are of interest, 'Yes, often have to, unfortunately; literature often inadequate', reflecting the observations made above, and 'Representative requested to assist if product unfamiliar . . .', which suggests that the trade representative is more likely to be contacted when there is a high level of uncertainty about the product. This may occur when the product is wanted for a situation not envisaged by the technical literature, but there must be sufficient information provided to suggest to the specifier that it is likely to be appropriate for the intended purpose. Enquiries will then be made by telephone, possibly followed up by a visit from the trade representative if this does not result in sufficient information being provided.

Office policy

The individual behaviour of specifiers will be affected by the policy of the office in handling marketing material and in its attitude to trade representatives. If the office filters the trade literature and presents a bar to trade representatives, specifiers may not be obtaining information that is otherwise available. Partners were asked about their relationship with the trade representatives. In the five very small offices, the partners declared themselves to be much more approachable, compared with those in the other offices. However, because these practices worked on smaller projects, the trade representatives paid them less attention. This was illustrated in one instance where a very small practice and a medium to large practice were located 100 m from each other in the same street. The very small practice was rarely visited by trade representatives, despite the fact that they had to walk past the door to reach the other practice. The larger practice complained about being pestered by the trade representatives, and the receptionists had been issued with instructions politely to deter them from visiting again: none who called without an appointment was entertained, and appointments were always to discuss products that related to a current project.

The experience of these two offices reflected a general pattern. The partners in the small offices said that they rarely saw trade representatives, while four of the five large offices had a policy of not seeing trade representatives unless they had been specifically invited into the practice by a member of staff. In general, trade representatives who called without an appointment were politely told by the office receptionist that no one was available to see them, or were 'palmed off' with the office junior or student architect. The only exception to this was one partner in a medium to large practice who made an attempt to see trade representatives when possible, partly because he was interested in new products and partly because he said it provided a break from an otherwise hectic job.

Another factor in this was the architects' views of the standard of trade representatives. All of the partners complained about and questioned their competence. When asked to explain their reason for this view, they said that many of the representatives had very little technical knowledge, often described as knowing less about the products they were trying to sell than the architects to whom they were trying to sell. However, these opinions were largely formed second hand from conversations with the specifiers in their office rather than from personal knowledge of the representatives, who they rarely met.

It had been assumed that the office receptionist would act as a gatekeeper to incoming telephone calls from manufacturers and to trade representatives when they attempted to visit the architect's office. While there was no evidence to the contrary, the only supporting evidence came from two of the receptionists who said that if trade representatives called without an appointment, the best they could hope for was a few minutes of the office junior's time, and noted that the representatives were 'rather a nuisance', seen as distracting them from more urgent matters. A receptionist in a small practice claimed that she dealt with the majority of telephone calls and trade representatives without reference to the partner or architects within the office because she knew when the staff did not want to be disturbed: she saw this as the duty of a professional receptionist and thought that all receptionists acted in this way. The partners were thus unaware of the extent of sales pressure.

In three cases where the reception areas were located on upper floors of buildings, a physical mechanism was employed in the form of an electronic entrance control system. Installed primarily for security reasons, it was also used by the receptionists to prevent people calling into the office without an appointment (trade representatives and contractors). All three of the receptionists said that they rarely let anyone into the office without an appointment (it was also very difficult to contact these three offices by telephone). In these three examples, the entrance control mechanism prevented trade representatives from calling at the office speculatively because they simply could not get into the building.

Marketing

Having identified a potential demand for their products, manufacturers need to exploit this by using the communication channels available to them. They need to create both interest and subsequent demand from potential adopters, but will market their products to different members of the building industry. For example, some products are designed to be sold only through builders' merchants, and others only to building designers, and this will inform their marketing strategy and hence those likely to become aware of the product. Their primary communication channels are trade literature and trade representatives, but the former covers a wide range of information formats. The range is from 'newsletters' and 'sales literature', which contain little or no technical information, to the more extensive 'technical literature', containing specifications and detailed drawings. It includes advertisements in the professional journals, direct mail and technical information, a growing proportion of which is also supplied in digital format via Internet downloads from the manufacturers' homepages. Trade literature is one way communication relying on the specifier becoming aware of the information and possibly making contact with the manufacturer to obtain further information before selection. The problem for manufacturers is that little of the direct mail will make it past the filtering mechanisms operated by offices.

This leaves journal advertisements as the other form of one way communication. Of the journals read in Britain, *Building Design* and the *Architects' Journal* were the most popular, both of which carry a limited amount of product advertisements and articles referring to products. *What's New in Building* and *Building Products*, both issued free of charge to specifiers, were popular; known as 'product journals', they primarily carry advertisements and articles relating to building products. The problem here is that advertisements in journals do not give many details. Certainly, specifiers are made aware of new products by this means, but they will not obtain any technical details unless the journal describes products in detail, although the existence of the product journals may mean that the significance of journal information may be higher in Britain than in countries where these do not exist.

The second method of communication is through the trade representative, either a 'sales representative' or a 'technical representative', a person who forms an interpersonal link between the manufacturer and the specifier's office. Although not employed by all building product manufacturers, they provide an important link between the manufacturer and the specifier's office. Rogers refers to such a person as a 'change agent', appropriate here because the manufacturer's intention is to change the specifier's behaviour:

> A change agent is an individual who influences clients' innovation decisions in a direction deemed desirable by a change agency. A change agent usually seeks to secure the adoption of new ideas . . . (Rogers, 2003: 366)

Rogers (2003: 369–370) has identified seven roles in the process by which an innovation is introduced to the potential adopter by change agents. Their roles are: (1) to develop a need for change, (2) to establish an information exchange relationship, (3) to diagnose problems, (4) to create an intent in the client to change, (5) to translate an intent into action, (6) to stabilize adoption and prevent discontinuance, and (7) to achieve a terminal relationship, i.e. develop self renewing behaviour. Observation of specifiers' behaviour shows that for trade representatives roles 1 and 4 are neither needed nor wanted. But to have any influence the trade representative must get beyond the office receptionist in order to see the potential specifier.

Specifiers have shown themselves resistant to innovations unless they already have a need created by a particular problem, in which case they will be actively looking. However, trade representatives seem to be unaware of this and continue to assume roles 1 and 4. But before they can carry these out, they have to get their foot through the door of the architect's office, and this is an important stage in the process. The trade representatives, or change agents, will attempt to communicate with potential specifiers in the architect's office, either by visiting the office or by telephone calls. Although they form an important link between the two different social systems, the professional office of the architect and the commercial world of the manufacturer, the very difference in their social position is a problem. Rogers has discussed the problem of compatibility between change agents and their clients, which he has called the heterophily gap. This is where the change agent is perceived as having low credibility by the potential adopters. In other diffusion research, difficulties in communication were related to different values and different levels of education, in Rogers examples, the change agents being more highly qualified than the clients. Unless a trade representative is also a qualified professional, it is the potential adopter who is likely to be the more highly qualified and who will certainly consider himself to be so. Thus, while the heterophily gap exists, the status levels are reversed compared with those normally found in diffusion studies. This has an effect that is different from that described by Rogers, where potential adopters sometimes felt reverential towards the change agent who was assumed to 'know better'.

A small study was carried out in which ten trade representatives were interviewed during visits to a specifier's office. The representatives were selected simply by the fact that one of the authors had a few minutes' spare time to see them over a four week period. Of the ten representatives interviewed, only one was a qualified architect, a strategy adopted by this firm because they were aware of the problem noted above. Four said that they had a qualification in marketing, while the other five claimed some experience of the construction sector. Those with a marketing background had previously sold a variety of products, from car components to chocolate bars. Although the interviews were not designed to be representative of all trade representatives, this clearly shows Rogers' heterophily gap. The representatives reported a reluctance on behalf of specifiers to look at their new product range, unless they happened to catch them when they were detailing and specifying buildings where such products may be of use to them. Clearly, the trade representatives are not unaware of the difficulty of their task, but it is probably not in the interest of those who have previously sold chocolate bars to draw this to the attention of their employers, and certainly not to explain the reasons to them (assuming they are aware of the reasons). This gives the impression that some manufacturers give little thought to their marketing strategy, with trade representatives employed more because it is customary to do so rather than with any clear purpose or strategy in mind.

How successful the trade representative may be at establishing a relationship with the architect's office depends on management strategies adopted by offices as much as on the marketing policies of the manufacturers, because offices are not simply passive recipients of the attentions of the manufacturers. Trade representatives may be regarded as necessary evils, but the evil attribute is as significant as the necessary and affects their ability to convey their message to specifiers. Nevertheless, as we shall see from one of the case studies, it is possible for a trade representative to build up a relationship with designers who regularly use his firm's products. In addition, the trade literature that comes into the office may be regarded as much as a nuisance as a welcome source of information. Because there are two kinds of message carrier, there are two different kinds of gate operating that need to be considered separately. They may be formal gates, consciously managed by the practice, or simply informal barriers.

A view from the trade representatives

A further five interviews were carried out with experienced trade representatives; each had been selling building products to specifiers for over ten years and therefore could offer an experienced view of the process from their perspective. The representatives were selected because they were good at getting through the gate, in contrast to many more who were not. These particular representatives were keen to use the meetings with architects for feedback on existing products and for views on new product development because they viewed themselves as an important link in terms of research and development of new building products. This may be a reflection of their companies' policies. All five complained that architects were very difficult to sell to, compared with contractors, but took it as the challenging part of their job. They were aware of the personal collections of trade literature used by architects, and all claimed various techniques for getting their information into these. They were aware of the 'foot through the door' problem that existed when visiting architectural practices, and noted that it was much less of a problem with contractors, QSs and planners.

To overcome the architects' resistance, the representatives marketed their products to other members of the building industry in an attempt to raise architects' awareness (and so have their products specified) by indirect means. Their trade literature and knowledge of their products were often communicated to the architects by pressure from, for example, the planners. They were aware of the influence of town planning officers over the choice of external materials, the influence of the QS where cost was paramount and the role of the contractor in changing products while the building was being constructed.

Three of the representatives explained the greater ease of communication with other professionals as due to arrogance on behalf of the architects, while the other two had concluded that architects were not interested in new products and so were immune to their sales techniques. The trade representatives felt that architects tried to put over an impression of being technically competent, when in fact they were often poorly informed about specific technical issues. This contrasts with the view of the architects, who were equally dismissive of trade representatives. In fairness to these trade representatives, they felt frustrated because they believed that a closer working relationship between the manufacturer and the architect would benefit both (this was not a view shared by the architects, who saw it as 'a waste of their valuable time'). Here, of course, we are comparing the views of a small number of competent representatives with designers' views of all trade representatives. The difference of view is not unexpected.

Four of the five representatives who marketed external materials said that it was just as important to get their information into the planner's palette of favourite products as into that of the designer. For example, the representative who marketed artificial stone products said that he had been particularly successful in selling a large quantity of his company's product directly through planners working in one particular local authority planning department, because they recommended his company's product by name to architects submitting schemes. Mackinder (1980) also found architects who reported that planners had a tendency to recommend products by brand name, a dubious practice from the point of view of both professional ethics and professional responsibility. The representatives also felt that they were successfully selling products to architects through the main contractors and, in some instances, the QSs, all of whom were easier to see and were perceived as more responsive to their products than the architectural offices.

Rogers (2003: 377) suggests that the greater the empathy between client and change agent, the greater the success in securing adoption. Some manufacturers employ architects as trade representatives in an attempt to reduce the heterophily gap. Ibstock was the first brick making company to do so (Cassell, 1990), a marketing strategy that was then copied by their competitors. This approach was reflected in their sales figures, with 60 per cent of all orders from architects' specifications, which their marketing manager attributed in part to their trade representatives' ability to 'get close to the specifier'. Rogers also noted that change agent success depends on the characteristics of the product innovation that the representative is attempting to have adopted. He commented that it depended on how well the innovation fitted into the recipients' existing belief system, and there is a parallel here in architecture. The representative trying to promote high quality facing bricks may stand a better chance of seeing the architect than if he or she was trying to promote, say nails, since the former is central to architects' interests, while the latter is not.

Rogers concluded that one of the most fundamental factors in the success of the change agent is the extent of change agent and client contact; the greater the face to face contact, the greater the likelihood of adoption and subsequent diffusion. Thus, the most important stage is to initiate contact with the potential specifier, i.e. the trade representative has to get a foot through the door. Some trade representatives may already have long established social relationships between one or a number of specifiers in architects' offices and rely upon those rather than relationships created for a single occasion. It is likely that these representatives will find it easier to introduce new products than those representatives who are trying to establish a new relationship with the office.

Because of the gatekeeping mechanisms in place, trade representatives can often only enter an office by invitation, i.e. once the specifier has already identified a need, and not before. As such, the trade representative may have limited impact as an agent of change. He or she may also have a limited role in raising awareness of 'new products'. However, once inside the specifier's office, there may be an opportunity to discuss other products produced by the manufacturer in the hope of specification at a future date.

Outside influences

It was noted above that manufacturers not only market directly to specifiers; they can target contractors and planning officers, so the influence of these on the specification process needs to be considered. There is the question of whether specific materials or components are requested by people external to the architect's office. The respondents said that the client was the most active in this (26 per cent recorded 'often' and 54 per cent 'sometimes'). Planners were said to be the next most frequently responsible for requesting particular materials or components, closely followed by the contractor. According to the sample, the QS was not regarded as having much influence (although this contradicts Mackinder's findings). However, it is not clear when such outside influences occur. The client may require certain materials to be used as part of the brief and the planning officer may have an early influence, but requests from QSs and contractors may be late in the process, made in an attempt to reduce costs.

Over half of the respondents said that they had changed a material, product or component on their last project as a direct result of a request by the contractor. This high proportion shows that the specification process continues while the job is on site and that changes may be made under pressure. The most common reason given for changing was non availability and/or unacceptable delivery times (to the contractor), although what evidence the architect may have for this is uncertain. While such changes have ramifications regarding the architectural office's liability, they also highlight the contractor's contribution to the process. Contractors appear to exploit time pressures as a way of influencing or forcing a decision that might otherwise not have been considered and so can form a barrier to building product innovations by leading to their discontinuance. Alternatively, they can introduce specifiers to products of which they were unaware.

Other reasons for changing products during the contract at the request of the contractor were because of cost and to suit the contractor's programme. There were also a few examples of change due to construction difficulties on site. Some comfort to those attempting to introduce new products through architects may be drawn from the respondents who confirmed that they had not deviated from the tendered scheme, best summed up by 'Change nothing if possible', reflecting the fear of liability, additional work and potential problems.

Product characteristics

The two characteristics of the products that affected choice were their cost and the relative newness of the product to the marketplace. Awareness of costs was shown to increase as the project progressed from scheme design through to detail design/specification. This is in line with expectations. Just under half of the sample (44 per cent) said that they were aware of price differentials between similar products, and a further 44 per cent said that they were occasionally aware of price differences, but we cannot be sure how early this might be in the process. When asked whether product cost would influence final selection, if they were aware of a range of product costs, 56 per cent said that it would often and 36 per cent occasionally. However, there is evidence of a reluctance on behalf of some manufacturers to disclose prices to the specifier, as indicated by the unprompted response that 'Suppliers and manufacturers [are] often reluctant to make current prices available'. Another unprompted comment was that 'No two products or subcontracts are ever the same', suggesting that it is difficult to compare like with like. These observations are in line with the experience of one of the authors. It is often extremely difficult to obtain prices from manufacturers, and those quoted to an architect may differ from those given to a contractor. While many architects may be aware of prices, it is not clear from the survey exactly how this information is obtained. In some cases, it may be from previous jobs.

Attitude to 'new' products

An important element of diffusion research is the length of time for a particular innovation to diffuse within a social system. Caution was emphasized in responses, with the majority waiting for over two years for a product to be on the market before adoption. Prompted comments ranged from 'Depends upon type of product, a brick has little to prove', to 'The longer the better – particularly if completely innovative', thus confirming the conservative behaviour reported by Mackinder.

Caution was noted when selecting products previously unknown to the specifier; 49 per cent occasionally select products of which they had no previous experience, but this depends to an (unknown) extent on the type of product, summed up by the comment 'wall ties no, wallpaper yes'. The majority of respondents indicated that they would wait until someone else had specified a product before specifying it themselves, indicating that either peer group approval or risk avoidance is an important factor in the adoption of innovations. Only a little over one third of the sample had specified products that were new to them or new to the market within the past year. Interpretation of the results here is difficult because perceived newness is not necessarily the same as the actual newness to the market. The list of materials recorded indicated that a product that is new to one specifier may not be to another, and furthermore, some of the products recorded as new to respondents had been available for a considerable period. There was no attempt to classify the respondents in terms of their speed to adopt innovations. However, this is an issue addressed in the diary of adoption (Chapter 9).

Only 31 per cent confessed to specifying products that were completely new to them, but an interesting list of products, components and materials was noted. One answer to this question was, 'Yes, too often', suggesting that problems had occurred. This was followed by a general comment; 'Use of innovative products is often restricted, not by any doubt on performance, but by insurance companies indirectly and contractors' unfamiliarity directly'. There is an interesting speculation possible here. Because control in France is exercised through the need to insure buildings and because the insurance companies, rather than the local authority, check the designs, there may be a greater resistance there to product innovations, and perhaps a greater reliance on Agrément certificates.

Attitude to risk

The relative avoidance of new products depends on an attitude to risk avoidance. Some architects' offices have a reputation for fashionable buildings that employ the latest technological advances in materials and building components; others retain their clients by acting in a conservative fashion, using the same type of materials on the majority of their buildings. However, the architect's office is engaged to take decisions on behalf of clients, and will have to accommodate their aspirations or limitations, cost control, local planning guidelines, etc., which can limit or enhance the innovative nature of decisions. The architect has a professional responsibility, a 'duty of care', to the client. Therefore, in situations where unfamiliar products are to be selected, extra care must be taken to ensure that the product is suitable for its intended use. In unfamiliar situations, new or untried products may be perceived as potentially dangerous, for if they fail, there may be an insurance claim against the architectural practice.

Regardless of its size or type, the office is liable for its actions and is usually protected by professional indemnity insurance. Research on the specification process assumed that the perceived risk associated with using building product innovations could act as a barrier to their adoption (Emmitt, 1997). Buildings are complex artefacts and the interaction of various components and materials comes to be understood only after a relatively long period in use. The specifier will have to decide (often at the preliminary design stage) whether to use traditional construction methods and products or risk using an innovative method. Cecil (1986) believes that a decision to adopt innovative materials, components or methods of construction presents a 'real enhancement of risk'. Not only does the specifier need to get it right, but the builder will also place greater reliance on the correctness of the specification, creating additional risk. Clearly, the threat of legal action against an architectural practice should be considered, and reasonable caution should be exercised.

Theoretically, the possession of professional indemnity insurance should allow the architectural practice the freedom to take risks and use unfamiliar products, but in practice, the office has to take reasonable precautions to prevent claims being made against it. The insurance company will raise the premium if a claim is made against the office's policy and may prohibit the office from using any products that have proved defective in the past. This may prevent or discourage selection of (experimentation with) unfamiliar products, despite the fact that the majority of architects' offices carry insurance cover; thus, the office norm may be to specify familiar products. Writing about legal liability, Hubbard (1995) used an example of a building where the roof leaked, and asked whether it was the architect's fault. In liability cases, defendants have to prove that they have taken appropriate action to discharge their responsibility. Therefore, if the architect

> had specified a recognized roofing system, and if her construction details had conformed to the accepted practice, then (in fair courts) she would have discharged her obligation and should be free of fault for the leak (Hubbard, 1995: 106).

The important words here are 'recognized roofing system', meaning one that other designers have specified, and 'accepted practice', meaning usual, or conservative, behaviour. There appears to be a difference between the product that is actually new to the market and the product that has been around for some time (thus has a track record) and has been specified by other design offices. A product perceived as an innovation by a specifier may be treated differently once he or she is aware that it has been available for some time, thus reducing the perceived risk of using it. However, it is important to recognize that it is the office that holds the insurance cover, not the individual. The perception of risk to the specifier in the office may be less than that of the partner.

Professional institutions offer advice and guidance to their members about a variety of matters, new product selection being one of them. For example, the RIBA's *Architect's Handbook of Practice Management* advises caution when considering the use of anything new, advising architects to evaluate carefully manufacturers' claims about their products' performance (RIBA, 1991).

In the questionnaire survey, loyalty of specifiers to favourite manufacturers was tested by asking whether, in circumstances where a manufacturer normally used does not produce the exact requirement, the architect attempted to find an alternative, compromise and specify the familiar manufacturer, or ask the manufacturer to revise the product. The majority, 74 per cent, would attempt to find an alternative manufacturer, so that brand loyalty is not particularly strong amongst specifiers: only 12 per cent said that they would compromise and stay with the familiar manufacturer. Nevertheless, office policy seems to favour known manufacturers when it comes to new products. Emmitt (2001) found that where manufacturers, and often their trade representatives, were known to the design office, partners were more relaxed in their attitude to information about new products. By way of contrast, they were very cautious when confronted by information from a manufacturer who was new to them. Some of the partners felt that they 'had a duty' to investigate new products but did not have the time to pursue it, while the others said that they did investigate new products, but only when the need arose. This observation supported the earlier work by Mackinder (1980), where one third of her sample noted that it was office policy to avoid the use of anything new if possible, preferring to stick to building products with which they were familiar. Her sample did, however, recognize that new materials and components needed to be monitored in case there were any advantages in terms of cost or performance.

Three of Emmitt's (2001) sample, all from very small offices, said that they would not use products that had been used by famous architects, because they were perceived to carry a greater risk of failure. In contrast, one partner from a small office said that it would influence his decision to investigate it further, while there was no strong view from the remainder. All of the sample said that they had to protect the excess limit on their professional indemnity insurance: the more claims the higher the excess, and so 'safe (familiar) products' were preferred. This was summed up by one partner, who said, 'We introduce new products to suit our, or our clients', circumstances not because a manufacturer wants us to. What's the rush?', a comment that supported the postal questionnaire respondents' desire to wait. It was this fear of unfamiliar products and manufacturers, perceived as risky, that had led to the employment of the gatekeeping mechanisms in the offices so that the literature did not reach specifiers in the design office. Architects seem to be taking the first line of Alexander Pope's dictum to heart:

> Be not the first by whom the new are tried,
> Nor yet the last to lay the old aside,

but whether they also believe in the second line is not clear. This reported resistance to 'new' products was also recorded in the observations reported in Chapters 9 and 10.

For anyone engaged in research it is rather depressing to realize that the results are not being used. For the authors who not only have contributed to this research, but have been and are practising professionals, and so users of trade information, it is doubly depressing to find that, although there has been a great deal of research that would aid more effective marketing of building products, there seems to be widespread ignorance of its findings. Only this can explain the continuing production of trade literature that does nothing except give employment to printers and fill waste bins, and the employment of trade representatives with poor marketing skills and insufficient technical knowledge. Manufacturers who have either made use of the research findings or simply developed sensible strategies through common sense will have a clear advantage over their rivals.

9 Monitoring the specification of buildings

As the adoption of products that are 'new' to the architect's office is likely to be a rare event in comparison to the large number of familiar building products that are used, a method was required that could trace a number of building product innovations from initial awareness to adoption (or rejection). Moreover, because memory recall is known to be unsatisfactory, some real time monitoring of the process was required. Participant observation was the method used.

The observations were conducted in the design office of a well established architectural practice that concentrated on industrial, commercial and retail developments. Approximately 90 per cent of their work was new build and 10 per cent refurbishment projects. The practice had gained a number of design awards, but had never been featured in the architectural press. It had a good reputation for service and delivery and had recently introduced a number of management techniques aimed at improving the delivery and quality of their buildings. The staff structure remained constant throughout the data collection period; three members of staff left and were replaced with similarly qualified people, discussed below.

The part of the drawing office in which the experiments were conducted was open plan with seven specifiers plus the author, comprising three qualified architects, three architectural technicians and an architectural student. The author (S.E.) was employed in a managerial role, and therefore, the tendency to ask 'why' certain decisions had been taken was a part of the daily routine and was unlikely to influence the behaviour of those being observed. An important aspect of the methodology was that the design office provided the opportunity to monitor both the adoption and non adoption of building product innovations as specifiers took decisions.

One of the difficulties in observing the design process is the time that this takes, although participant observation has been used by researchers such as Cuff (1991). This was also possible here, and a diary of adoption was kept over a forty month period. During this time, an attempt was made to record any event that affected the process of adoption. One of the problems, however, was to ensure that this was a practical proposition. To ensure this, nine products were selected for monitoring and the first stage of the process, that is, the initial knowledge stage, was deliberately started by the observer. The experiment was begun by requesting trade information about a number of products by using the 'reader reply' card enclosed with one of the free 'product' journals received by the office. Products were selected to cover a variety of uses:

- product 1: a uPVC product to be used with coping bricks for brick walls
- product 2: an external decorative facing panel
- product 3: a uPVC cavity closing system incorporating a thermal barrier
- product 4: reinforcing mesh for retaining soil on sloping sites
- product 5: a timber flooring system for use as an internal finish
- product 6: a product to assist the pouring of concrete on site
- product 7: thermal insulating building blocks
- product 8: an internal partition walling system
- product 9: an electronically controlled door lock.

There was standard office procedure, in which trade literature received by the office was passed from the senior partner to the technical partner on a daily basis and circulated to the staff at approximately four week intervals, before it was filed in the office library. The technical partner's job was to assess the literature and only circulate that which he thought would be both useful and technically acceptable, i.e. he acted as a technical gatekeeper, controlling the flow of trade literature to the specifiers in the office. To enable the observer to monitor the effect of this, the receptionist (who opened the mail before passing it to the senior partner each morning) was asked to record the date when information about the nine products was received in the office.

Trade literature relating to all nine building product innovations was received over a two week period, but the only literature to reach the technical partner related to products 1, 2 and 3. Two thirds of the information requested had failed to get past the senior partner during the morning ritual of opening the mail; it had been thrown away. This posed a slight problem for running the observation.

The significance of this filtering process is discussed below, but it hindered the research by reducing the sample to only three products. Therefore, the information was requested again by telephoning the manufacturers directly. Since it was likely that the senior partner would again throw the information away, the receptionist was asked to intercept it and pass it directly to the technical partner, thus bypassing the senior partner. Information on five of the six building products was received and placed in the technical partner's in tray by the receptionist. Information relating to product 9 was never received. Thus, after five weeks, trade literature relating to eight of the nine building products had been received.

Two gates

Earlier, the filtering of trade literature was discussed, and this observation helps to illustrate its complexity, confirming evidence of more than one gate through which technical literature has to pass before it is retained within the office library of information.

The first gate

The senior partner, when asked about his gatekeeping actions, could not recall the particular examples, but did confirm that 'the majority' of trade literature received through the post every day was thrown away. He said that his decision to keep or reject information was 'something done quickly and based on experience'. Literature that looked as if it was well produced and might be of use to the office tended to be passed to the technical partner; the remainder (approximately two thirds) was thrown away. This reduced the physical amount of literature passed on to the technical partner and the amount of time he spent assessing it. The technical partner's job was to limit the office's exposure to risk by controlling products available to the specifiers in the office through this means. Thus, a product perceived as 'risky' should be rejected before reaching either specifiers or the office library. Although the senior partner delegated the task of vetting trade information to the technical partner (as described in the office manual), he acted as a gatekeeper to reduce information overload (which was not mentioned in the manual) before it reached the second gatekeeper.

The second gate

At the start of the sixth week, the technical partner circulated a tray containing his approved trade literature to the specifiers in the office. This contained information from twenty seven different manufacturers, ranging from a four page brochure on handrails to three binders containing a variety of products. Six of the brochures related to the preselected sample, so the technical partner had rejected two of them based on his assessment of the trade literature. These were products 3 and 6.

When asked about this process, he said that product 6 was of interest to him because he thought that it would save time in construction, and he had requested additional technical information by telephone. Curiously, the information was placed in his personal collection of literature and not circulated to the staff or filed in the office library. Despite his personal interest in this product, it was never specified by the practice and was still in his personal collection of literature at the end of the experiment. When questioned about this, the technical partner said that he fully intended to use the product, but had not had the time or the opportunity to investigate it. Having said that, he still thought it was 'a good product with potential and would use it at some point'. The important point here is that he had, unwittingly, prevented the specifiers in the office from seeing the information and, therefore, it could not be specified by them unless they became aware of it from some other source.

The second rejected product, product 3, provided an example of a product that, as far as the technical partner was concerned, was launched too early. It had been launched to anticipate the introduction of the revised Building Regulations to reduce the effect of thermal bridging in cavity wall construction at window and door openings. Although the technical partner thought that it was an 'interesting product', he was concerned about its use because it was made of uPVC. Because he thought it 'too risky to use', the information was thrown away. Despite this, the product reappeared during the monitoring period (see below).

Six products were perceived as being potentially useful and, more importantly, 'safe'. Although product 1 was also made from uPVC, it was not viewed with the same suspicion as product 3 because it was for use on boundary walls rather than cavity walls of a building. This was regarded as less of a risk to the architect's office by the technical partner.

Information about the six building product innovations that had passed through the second gate was circulated to the potential specifiers in the office for a week and then placed in the library for filing. Of the six, products 5, 7 and 8 were never considered for specification during the monitoring period, despite the fact that projects were being designed and detailed where similar products were specified. Information about these products was still in the office library at the end of the experiment, presumably by then out of date.

Comparison with the postal questionnaire

A question in the postal questionnaire was designed to measure the awareness of the same nine products by specifiers in other architects' offices in the same region. The questionnaire was posted twenty five weeks into the monitoring period, and replies were received between four and eight weeks later. For the purposes of making a comparison with the diary of adoption, it was assumed that they were completed during week 28. The question asked was: 'The following products have been launched onto the market within the past 12 months, would you please indicate those you are aware of, those you have considered using and those actually specified'. The replies are shown as the number of respondents from a total of 138. The figures are compared with the diary of adoption, where the specifier's awareness use is indicated (Table 9.1).

Fifteen weeks into the observation, product 1 had been specified by the office and recorded both the highest awareness and highest level of specification by respondents to the postal questionnaire. Product 2 had also been specified by the diary office, recorded the second highest awareness in the questionnaire,

Table 9.1 Comparison of diary office and postal questionnaire respondents

Product	A	C	S	Diary office
Product 1	69	15	5	Specified
Product 2	66	17	3	Specified
Product 3	25	8	2	Unaware
Product 4	31	3	2	Aware
Product 5	37	2	4	Aware
Product 6	19	0	0	Unaware
Product 7	48	6	1	Aware
Product 8	24	13	1	Aware
Product 9	18	4	0	Unaware

Twenty of the 138 respondents indicated that they were unaware of any of the products. A: awareness; C: considered; S: specified.

had been considered by fifteen respondents and had been specified by three of them. Of the four building product innovations with the lowest awareness in the postal questionnaire, three of these had not made it past the two gatekeepers in the diary office. This may be coincidental, it may be a reflection of the marketing strategy employed by the company, or they may have met with similar resistance in other offices. Whether this is poor advertising or a poor product is difficult to say. Nevertheless, the similarities between the postal questionnaire and the architect's office in which the monitoring was taking place suggest that the actions monitored were representative of behaviour in other offices.

The specification of 'new' building products

The events recorded during the observation period are presented here under the product number.

Product 1

This manufacturer's trade representative had been invited into the office to discuss the possibility of using their bricks for a particular project. During the meeting (with one of the authors) in week 1, the representative took the opportunity to introduce the coping system (product 1) and left two copies of literature, one for the office library and one for personal use (i.e. for inclusion in a palette of products). The information was placed in the technical partner's in tray. The introduction of this product was very timely, since it appeared to solve problems that the practice was having with vandalism to boundary walls (traditional detail) on several inner city sites. A week after his visit, the trade representative delivered a sample of the product (a 'tactile demonstrator') to the office, and this too was placed on the technical partner's desk. It also attracted considerable interest from the specifiers in the office. This well known manufacturer believed that if a specifier can actually handle a product, the chances of specification are much higher because the product can be explored in greater detail than paper information allows. Their trade representative also offered to take specifiers to their brickworks to see a demonstration of the product being built into a wall. This was declined because the tactile demonstrator served a similar function.

One of the specifiers specified product 1 as a trial on a boundary wall for a new project during week 3. The use of trials has been noted by Rogers, so this accords with his observations. His decision to adopt this product as a substitute for a traditional tile creasing detail was made after checking technical queries with the manufacturer's technical department by telephone. Approval was also required, and given by the technical partner, before seeking approval from the client, which was also granted.

The product was not delivered to site as programmed because of manufacturing difficulties at this early stage in the development of the product. This is an unusual situation because most of the innovations considered here are only innovations in that they were new to particular specifiers. This product was also an innovation in the commonly used sense of the word, i.e. it was a new product. Teething problems with new products are an additional factor that, in some cases, can lead to rejection or discontinuance of the product, but it has not been possible to explore this effect. The product was eventually delivered to site two weeks late. Although this led to a request from the contractor to change to a traditional detail, to save time, this was declined by the specifier. This is an example of pressure to change products because of time pressures associated with the building programme.

The client for this building made frequent visits to site and, having seen the coping system, said that he liked it and asked the technical partner to use it on all future projects. Standard detail drawings for this particular client were updated to include product 1 on boundary wall details, thus ensuring automatic specification on all future jobs. However, standard details for other clients in the office were not revised because the technical partner did not feel it necessary. The adoption of some innovations may be client specific, suggesting that offices may have other files of approved products for repeat use with regular clients that may differ from the normal office palette.

During week 51, the technical partner was telephoned by a specifier working in an architect's office elsewhere in the country who also worked for the same client. They had been asked by the client to specify product 1 on all of their future projects and had telephoned to seek further advice (and peer group approval), since the product was new to them: this led to the other office adopting this product as a standard detail. The two architectural offices had an informal relationship because they shared the same client, which meant that technical knowledge was occasionally shared.

Evidence of potential discontinuance came during week 55, when the quantity surveyor (QS) suggested that the specifier should revert to a traditional boundary wall detail to save money (on a project that was running over budget). By this time, the product had been specified on seven projects because it had become a standard detail for the office. The request was declined, but the issue of cost was picked up by the client, who then asked the practice only to use the product on future projects where absolutely necessary. Thus, it was only used as a standard detail where vandalism was thought to be a problem. During the forty month period, the building product innovation was adopted quickly on projects for a particular client, but not for other projects. Pressure to change came from the QS initially, then from the client.

Product 2

During week 8 of the monitoring period, one of the specifiers had to redesign a building to suit the request of a planning officer, who had refused to accept the proposed cladding material, timber boarding, to the gable walls of a building. The material had been annotated on the drawings submitted for full planning approval, and the planning officer had stated that he would recommend that the application be refused unless the material was changed to 'something better'. There was evidence here of contribution to the specification process from someone outside the immediate social structure of the project, in this case exerting pressure to change the material already selected by the specifier.

It was standard office policy to report any changes in design and the reasons for them to the client. In this case, the client (the same as above) asked the specifier to resist the requested change. This resulted in a number of telephone conversations and an unsuccessful meeting with the planning officer. At the meeting, the planning officer said that he would accept brickwork or render. Brickwork was rejected by the specifier because it would have required changes to the design of the structure of the building but, more importantly, would have required the revision of a number of drawings and the preparation of two new detail drawings. The use of render was also rejected by the specifier because of the anticipated maintenance costs of the material. A material was needed that could be substituted for the timber boarding without the need to alter too many drawings.

Information about such a material, product 2, had recently been circulated to specifiers in the office and placed in the office library, and the specifier went there to look for suitable products and took out this information. Information about similar products was also available in the library, but was ignored by the specifier because product 2 was fresh in his mind (he had remembered its being circulated in the office). The information about similar products had been in the library for some time, which may have influenced the decision to ignore it. This incident highlights the value to the manufacturer of a specifier becoming aware of an innovation during or just before his need for it. Of course, had this information not been so recently circulated, the specifier might have paid more attention to the other options.

He telephoned the company to request samples of the product and for a trade representative to visit the office to discuss its use in more detail, i.e. he was seeking reinforcement. The representative visited the office the following week and satisfied the specifier's questions about the cutting and the fixing of the product on site. Guarantees and a technical specification were provided by the company and checked with the technical partner, who approved its use and who also obtained the client's approval. Drawings indicating the use of product 2 and a sample of the product were taken to a meeting with the planner the next week and were approved. Thus, the building product innovation had been adopted by the office over a three week period. The specifier placed a separate copy of the trade literature (which had been brought to the office by the trade representative) in his personal collection and, when questioned about this, he said that it was so that he could refer to it easily if there were any queries from the building site, rather than having to get the information out of the library. He also liked it and intended to use it on future projects.

During week 33, the product was delivered to site and, following a site visit, the specifier was overheard telling a colleague in the office that it was a good product. However, after the product had been fixed on site, the client said that he did not like the finished appearance. He preferred the timber boarding and said that this product should not be used again on any of his buildings. Although the building product had been adopted by the specifier, there was no opportunity to specify it again until week 149 (no one else in the office had specified it either). It was for the same client and, once again, a planning officer in a different planning authority said that timber would not be approved. The client stated quite strongly that '. . . on no account was the practice to use this product'. There was nothing wrong technically with it: he merely did not like its appearance. As a consequence, it was not specified, and a traditional render detail was used (despite the concern over maintenance costs). Despite this, the specifier retained information about this product in his palette.

Product 3

This was the product that the technical partner had initially rejected because he perceived it as too risky to use. Following his initial rejection, requests were made by a number of building contractors to the technical partner for the office to specify a simpler and quicker system of closing the cavities at openings in the wall. Both product 3 and a competing system were suggested by three different contractors over a twelve month period. In this case the product had already been used by the building contractors (specified by other architectural offices), who found the product both easy and quick to fix on site. The contractors had found that the product had a high relative advantage, was compatible with existing construction details and was simple to understand. Another important point was that they had been able to use the product on site without risk to themselves (because it had been specified by an architectural office who therefore carried responsibility for its selection). In addition, two of the contractors had adopted the product as a standard detail for projects on which they were responsible for specification (i.e. design and build projects and speculative housing developments). It was the contractor's experience of using the product that was influential in the technical partner's decision to reassess the information. This also confirms an observation made by Rogers that some adopters will wait until they have seen an innovation successfully used by others before adopting it themselves.

In this example, knowledge about the innovation was communicated to the potential specifier by a source other than the manufacturer, through interpersonal communication between specifier and contractor, which was also influential in reducing the technical partner's concern about the innovation. The product innovation had two advantages: it offered improved thermal insulation (improved performance), but was initially rejected on the grounds of offering insufficient advantage versus risk, and it offered ease of installation on the building site (improved buildability), and was eventually accepted by the specifier at the request of the contractor. The specifier was concerned with the first characteristic, while the builder was concerned with the second.

Technical information and a sample of the product were requested by telephone directly from the manufacture and sent to the office by post. At that time, the product had been on the market for about two years, and the company offered names of other architectural offices that had used it in an attempt to endorse the product. Information about the other, competing, building product innovation suggested by the contractors was ignored by the technical partner, simply because he had seen both products on site and favoured product 3.

The technical partner checked the product's technical specification, from the literature, and then asked one of the specifiers in the office to revise the working drawings for one project, which was carried out during week 70. Thus, the product innovation had been adopted on a trial basis and none of the other jobs in the office was altered at this time. The product was fixed on site during weeks 96–98. There were no problems with either delivery or fixing, and during week 118, the technical partner issued an A4 sized drawing, together with an internal memorandum, which asked all staff to revise the detail on all future jobs to include product 3. It had been adopted by everyone in the office through what Rogers describes as an authority decision. At the end of the monitoring period, the product was still specified on the standard construction details for all new build work.

Product 4

Information about product 4 was circulated in the office and then placed in the library. No interest in the product was recorded until week 68, when a structural engineer suggested its use during a telephone conversation with a specifier, in which they were trying to find a solution to a design problem. A photocopy of the technical information was sent from the structural engineer to the specifier that was perceived as an innovation (he had forgotten that he had seen the literature earlier). In this example, the structural engineer contributed to the specification process by providing knowledge about the product to the specifier: awareness was not directly from the manufacturer but through a third party, an external contributor.

Further technical information was requested by the specifier (it was already in the library, but he did not go and look), and a trade representative delivered this and a sample of the product to the office during week 71. A decision was taken to use this. The structural engineers had used it previously when working with other architects' offices, and information was sent to the QS for cost appraisal. At week 74, the QS reported that its inclusion would put the estimated cost of the project over budget. After discussion, a revised scheme was developed by the structural engineer and the specifier that did not require the use of this product. Although it was not adopted in this case, information about it was added to the specifier's palette of favourite products for future use (postponed adoption). It was not used during the monitoring period and was still in his collection of personal literature at the end of the experiment, suggesting that he may use it were a suitable situation to occur.

Contributing factors

During the monitoring period, it became apparent that several contributing factors were present that were not related to any specific building product innovation but were important in relation to the application of the Rogers model to the building industry, and it is important to note these. Some general observations can be made about the behaviour of specifiers in the office, and there was some influence on the events because of the movement of staff that occurs between offices. The contribution of local authority town planning officers also requires some discussion.

The specifiers' behaviour

During the experiment, the practice received certification for its quality assurance scheme. This should have affected the specifiers in the office because there was a clause that theoretically stopped individual members of staff retaining their own source of literature. This was to prevent the retention of out of date material and, according to the quality manual, they could only use information from the office library. In fact, what happened was that individual members of staff simply kept their personal collections of literature away from public view, taking them off open shelves and locking them in the bottom drawer of their desks. By this means, they continued to use them, so continuing the tendency to use the individual palette before looking elsewhere. One may well ask whether strict enforcement of the quality manual would have brought specifiers into contact with a greater range of trade literature and increased their knowledge base and so possibly the number of innovations considered. However, the effect is likely to be small if one assumes that specifiers mostly look up details of products that they have used before.

Staff movement

During the monitoring period, there was an economic recession that seriously affected the building industry. This meant that there was a tendency for staff to stay in their current employment (unless made redundant), rather than to move jobs frequently, as had happened in the economic boom of the late 1980s. The office in which the data collection was carried out, unlike the majority of other architects' offices at the time, did not make anyone redundant because of the recession. This meant that the office was a relatively stable environment in which to carry out the experiments, although three of the specifiers did leave the office to take up employment elsewhere and were replaced with three new staff with similar qualifications. This provided an opportunity to monitor any information transfer between offices through the movement of staff.

The first to leave (week 63) was replaced by someone of a similar age and experience who had been working in a smaller architect's office in a nearby town. The outgoing member of staff took his personal collection of literature with him, while the new member of staff brought his to the office (despite the restrictions described above). The specifier who had left the practice was interviewed twelve months after his move in an attempt to see whether he had introduced any product innovations to his new office. His new office had adopted products 1 and 2 as approved products before he took up his new job, and he had specified both (although he had not used product 1 while in the office being monitored). He was unable to confirm during the interview whether or not he had introduced any new products to his new office. The other two who left took up jobs that took them away from product specification.

No attempt was made to analyse the content of specifiers' personal collections of literature and there was no evidence that the three new members of staff introduced any building product innovations to the office being monitored. All three new members did, however, specify products 1 and 2 because they had been adopted by the office and were included in standard details and the master specification for certain clients.

Contribution of planners

During the experiment, there was resistance to the use of brick on a building project. When the planning drawings were submitted for approval, the planning officer asked the specifier to change to artificial stone. Following a number of lengthy discussions with the planning officer, she finally granted approval for the use of brick. In this instance, the pressure from a planning officer exerted on the specifier to change materials would have led indirectly to the rejection of product 1 on this particular project; however, two other factors emerged.

The artificial stone manufacturer's trade representative had spent a lot of time convincing the planning officers that his company's product was worthy of consideration; he had seen all of the planning officers in areas where stone had commonly been used in the past, and had placed trade information and product samples in the office library of the planning departments. Clearly, the representative had done an excellent job because the planning department was legally outside its remit in insisting on the use of this company's artificial stone products: it could recommend the use of a stone treatment, but not a particular company's product. But significant here is that awareness of a building product had come from the planning officer, an external contributor.

Secondly, when the town planner was interviewed, it was found that in addition to the planning department's extensive library of trade literature and samples, mainly external materials, the planning officer had her own personal collection of favourite materials. Clearly, the manufacturers and their trade representatives were aware of the planners' role in product selection and saw it as another route to raising specifiers' awareness, albeit indirectly.

Implications

This case study helps to highlight the complexity of specifying buildings while also identifying a number of issues that must be addressed in the pursuit of design excellence.

Because the monitoring took place over a long period, a number of events was recorded that might have been missed in a shorter period, such as examples of discontinuance (product 1) and adoption after initial rejection (product 3). This helps to reinforce the importance of longitudinal data collection exercises when researching the diffusion of innovations. The diary of adoption highlighted the complexity of the specification process, in particular the contribution made by people from outside the architect's office. This demonstrated how information about building product innovations can enter the architect's office through interpersonal communication channels generated by the building project itself, rather than directly from the manufacturer, a process that bypasses any formal gatekeeping mechanisms.

The filtering of trade literature as it entered the office was formally controlled by two gatekeepers, and evidence of this supported the gatekeeping research reported earlier. However, despite the existence of a quality management system that had regulations supposedly ensuring that all trade literature in the office was up to date, the specifiers kept their own collections out of sight of the office quality manager. Some of this had been brought from other offices, and some of it must have been out of date.

The specifier's palette of products was used to store information about products that had been adopted and also to store information about products that had been investigated but, for whatever reason, were not specified. This was so that they may be considered for use in the future, this being referred to as postponed adoption. Furthermore, there was evidence that the external contributors, certainly the planning officers, also kept a palette of favourite products.

Awareness of new products was clearly linked to need. When a specifier had a need, the building product innovation was adopted quickly, as was the case with products 1 and 2. Where there was no immediate need, there was a tendency to ignore information about the innovation or forget about it (e.g. product 4).

Two of the building product innovations, products 1 and 3, were both made of uPVC, a material disliked by the technical partner, and both used in conjunction with brickwork. So why did he initially adopt one and reject the other? Product 1 was manufactured by a company known to the office and hence was viewed as carrying little risk. Product 3 was a new name to the office and was perceived as carrying a greater risk because the manufacturer was unfamiliar. This shows that the track record of the company promoting the building-product innovation was an important factor at the awareness stage for this individual, an observation that contradicts the views recorded in the postal questionnaire. There was no evidence that the cost of a particular product was considered by the specifier; this was left to the QS and only addressed by the specifier when identified as a problem. Clearly, the manner in which the specifier's office is managed, the personality of the individual specifiers, as well as the influence of others party to the construction process will influence the uptake of new products. So, too, will the characteristics of individual products.

10 Observing specifiers

Some discrepancies were found between the findings of the questionnaire and those of the diary of adoption. In some respects, the way in which people believe that they behave does not seem to be the way in which they actually behave in practice, and this has also been shown by the closer focus provided by the diary (Chapter 9). The interviews that were carried out also showed the varying effects of office policy, in particular how difficult it might be for some products to get past the initial gatekeeping procedures. But the investigation also showed mechanisms by which these might be bypassed. It showed that the adoption or rejection of building products might depend on the very particular circumstances of individual jobs. That being so, it is going to depend on the behaviour of the individuals involved in the process.

In circumstances where the specifier is unable to use his or her palette of known products and seeks alternatives, it is sometimes a matter of choice between competing innovations. This is not something that has been considered by the Rogers model, but which is useful to explore how the basic model of diffusion needs to be adapted for the design situation. The question is, how does the behaviour of the various players affect the adoption or otherwise of an innovation? This chapter focuses on the behaviour of individual specifiers using a case study approach. The three case studies are concerned with the factors that cause a specifier to look beyond his palette to building product innovations and the ensuing innovation decision process.

The specifiers observed in each case study were all male and were working in different design offices. All three offices were of a similar size and worked with a broad range of building types. The first case study was primarily concerned with observing a situation or an event that caused the specifier to look for building product innovations and was conducted in the same office as the research reported in Chapter 9. The second case study looked at a specific factor that caused specifiers to consider building product innovations, the implementation of new Building Regulations relating to the thermal performance of buildings. In the third case study the emphasis was on the detailing of building designs and the relationship to the specification of building product innovations. Combined, these observations provide an insight into the behaviour of five specifiers as well as raising issues of relevance to design managers.

Case study 1: Specifier A

The objective was to observe a situation or event that caused a specifier to investigate 'new' products; therefore, the specifier(s) would be self selecting and the period of observation determined by the individual and the project life cycle. There was a danger that the critical event might not have occurred when the observer was present. What was significant was that the opportunity for observation did not occur for a considerable amount of time because specifiers in the office continued to select products that were familiar to them, thus demonstrating that the adoption of building product innovations is a rare event, supporting the views of Mackinder's (1980) sample. Eventually, a situation arose that could be observed and recorded for the duration of the process. The specifier observed sat next to the author and had the unusual habit of talking out loud while he was working; although this irritated other members of the office, it made for an ideal subject since the thinking process and the decision-making process, usually hidden from an observer, were quite transparent. The observations were recorded by the author in writing in a desk diary, recording both the actions observed and the length of time taken, and then analysed. At the end of the observation process the specifier was told that he had been observed, and his consent was obtained both to use the material gathered and to interview him to explore the motives behind his decisions.

The material gathered was a detailed account of a specifier and the influence of his direct surroundings. The specification act was highly interrelated to other activities within the office, and the process reported below had to be separated out from other tasks that engaged the specifier during his working day (working on two other projects concurrently, attending meetings, dealing with telephone calls, site visits, etc.). The observation reported below helps to illustrate the complexity of the decision making process that occurs during the specification process, as illustrated in Fig. 10.1.

The specifier had two weeks to produce the working drawings for four, single-storey, retail units. Three of the four were to be built with timber rafters and concrete interlocking roof tiles, and the fourth was to be detailed with a structural metal tray and a profiled steel roof to suit a particular client's requirements. The designer first detailed the three units with the tiled roof, a task carried out quickly because he was familiar with this form of construction. He had used a very similar roof construction on a previous project, and the drawings produced for it were used to gain information for use here, thus reinforcing the tendency to use familiar products, observed in earlier research (Mackinder, 1980). When he attempted

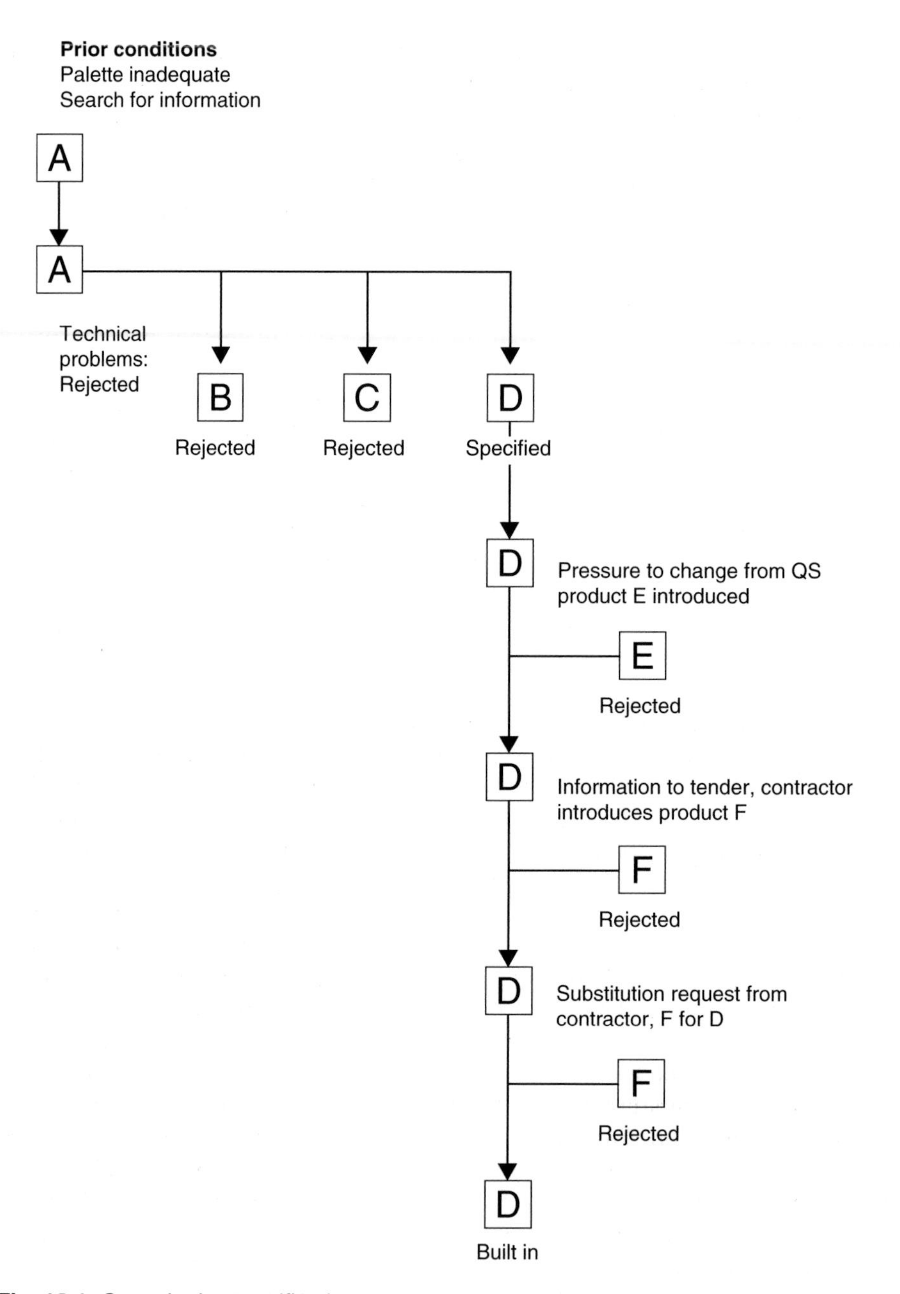

Fig. 10.1 *Steps in the specification process*

to detail the metal roof, a form of construction that was unfamiliar to him, he was unable to draw on his previous experience because he had not worked on any similar projects, although other specifiers in the office had. Since his personal collection of literature did not contain any information that could help him, he was forced to look for products that might solve this particular need: he was forced to search for information about building products that were new to him (consistent with stage 1 of the Rogers model).

Innovation A

His first action was to ask other specifiers in the office whether they had experience of detailing such a roof, i.e. he first sought knowledge from his peers, drawing on the collective experience of the office, an action he later said was taken to 'save time looking in the library'. A colleague suggested a product that the office had used successfully before, product A, but which was new to the specifier. He spent approximately ten minutes talking to his colleague to gain more information and to establish whether or not the product was suitable for his particular requirements. He was seeking to reduce his level of uncertainty, which is consistent with the persuasion stage of the model (stage 2).

Satisfied with the information gathered, he then sought further information about the product from the office library to enable him to make a decision. Because the trade literature was not comprehensive enough to solve all of his queries, he telephoned the manufacturer to request additional literature. The manufacturer offered to send a trade representative, the change agent in Rogers' terms, to the office to assist with any queries. This was declined by the specifier, who later said that he did not have sufficient time to see the representative. Information was received by post three days after the request (during which time the specifier had been working on another project). After reading the information, he made a decision to specify product A and continued with his detail design work. The innovation had been adopted, consistent with stage 3 of the model 'decision', but his decision was subsequently revised.

While detailing the roof, he discovered a technical problem that he could not resolve from the literature, so he telephoned the manufacturer's technical department for clarification. During the telephone conversation, it became clear that product A would have to be modified to resolve his particular problem, but the manufacturer did not have a 'standard solution' simply because they had not considered such a possibility arising. This resulted in a state of dissatisfaction on behalf of the specifier, who immediately went to the office library to search out an alternative. He did not use the electronic database or the printed product compendia; he selected entirely from the trade literature on the library shelf, a search pattern that the specifier later confirmed to be the quickest way of finding suitable products, further emphasizing the time pressures exerted during the detail design stage.

Innovations B, C and D

From his search in the library, a further three building product innovations were selected, a task on which he only spent ten minutes. The library contained trade literature from ten manufacturers of similar metal roofing products, seven of which were rejected simply because their technical details in the literature was seen to be of 'poor quality' by the specifier. On returning to his workstation, the specifier telephoned all three manufacturers to question them about their products. Product B was discounted because the technical representative was perceived as not knowing his product well enough (described as a 'complete idiot' by the specifier). Product C was rejected because the company would only answer technical queries by sending a trade representative to the office; since the earliest appointment would be too late for the specifier to complete his task to programme, the product was rejected. Product D was adopted because the technical representative 'knew his stuff' and had offered some additional practical advice to the specifier that helped him to complete his detailing quickly. In line with office policy, the specifier went and spoke to the organization's technical partner, who was responsible for granting approval for the use of any product that was new to the office. Following a short discussion, approval was granted, and the specifier returned to his desk to resume his detailing of the roof. Product D was referred to by name on the drawings and later in the accompanying written specification, a task he completed within the two week programme.

Pressure to change: innovations E and F

The production information was then sent to the quantity surveyor (QS) for production of the bills of quantities and also for a cost check of the design against the original budget. During the three week period taken to complete this task, the QS telephoned the specifier to suggest an alternative to that specified in order to save money. The alternative, product E, was unknown to the specifier, so it constituted a further building product innovation, one introduced to him by a contributor to the design process who was primarily concerned with the cost of the product, not its technical performance. This illustrated the contribution made from outside the design office during the specification process, with pressure to change the specified product, and also the introduction of a product that was known to the QS but not the design office.

Product E was immediately rejected by the specifier simply because he had invested a lot of time in solving a particular problem and did not want to go through the process again with a different product and different fixing details. He made no attempt to analyse the information, despite the potential cost savings reported by the QS. As a result, product D survived this first attempt at specification substitution and was included in the documentation sent out to competitive tender. This illustrated the contribution made from outside the office during the specification process, with pressure to change the specified product, and also the introduction of knowledge about additional product innovations, confirming one of the stages of the theoretical model, and also again illustrates the effect of time on the process. This time constraint is specific to the design process and not part of the Rogers model. Time pressures appeared to be of paramount importance to the specifier.

Evidence of specification substitution

The lowest tender was accepted by the design office and approved by the client. But the contractor had also submitted an alternative, lower, contract sum based on a revised specification. Twenty three products had been identified for which there were cheaper alternatives, ranging from the facing bricks and cavity insulation to the ironmongery for the internal doors, and including one to replace the steel roofing system, product D. Thus, a further innovation had been introduced, product F, on the contractor's list of suggested alternatives. Individual product costs were not listed; instead, a total cost saving had been identified, compared with the original products specified. The client asked the specifier to analyse the alternative tender sum and then advise him which to accept. The proposed alternative, product F, was unknown to both the specifier and the office.

Although the specifier wanted to reject the substituted products immediately, further information had to be sought so that a report could be made to the client. He telephoned the manufacturers and asked a number of questions about delivery and guarantees. The answers raised further issues to be investigated, and since he did not have the time to pursue them, he rejected all of the substituted products, including product F, recommending to the client that the cheaper products were of insufficient quality and/or not acceptable visually (e.g. the facing bricks). This whole process, including the writing and faxing of the report to the client for approval, was dealt with in twenty four hours. The contractor was appointed the following day with no change to the original contract documents. Product D had survived.

Further attempts to change a number of products, including product D, were made by the contractor after the project had started on site on a number of occasions during the thirty week programme. This confirmed evidence of specification substitution reported in earlier work, although in the event, the specifier refused all requests. First, the contractor claimed that the specified product could not be delivered to suit the programme and proposed product F again. This was found to be untrue when the specifier checked with the manufacturer, who confirmed that the contractor had made no attempt to place an order for product D (presumably, the contractor had hoped that the specifier would accede to his wishes without checking). The request was refused. After the first request had failed, the contractor again proposed that product F be substituted to save money (for whom was never made clear). Again, this was refused by a specifier keen to see his design decision transferred from drawing board to finished building. Eventually, product D was delivered to site, to programme, and built into the building without any problems being reported from site. Thus, after a number of attempts to change it, it had finally been implemented (stage 4 of the model).

Toward the end of the project, the specifier added product D to his personal collection of literature for use at a future date. It had now become part of his personal inventory of products. This could be seen as evidence of the confirmation stage (stage 5) because the likelihood of the product being used again is high, although not observed in this study.

The persistence of the contractor deserves some comment here. Although no enquiries were made to explore the reasons for this, there are several clear possibilities. The most likely is that product F was one that the contractor had used before and therefore presented no problems for him. If he was unfamiliar with the specified product, he might be uncertain about his ability to use it without difficulty. Of course, it is also possible that he was only too familiar with the specified product and had previously experienced some difficulty in using it. Another possibility is that product F was available through the contractor's normal supplier and the amount of discount was greater than that for the product that was not so available; thus, financial considerations affecting the behaviour of the contractor cannot be ruled out.

The specifier's viewpoint

When the observation period was complete, the specifier was interviewed to address questions about the process of specification and his attitude towards building product innovations. He was asked to recount the actions he went through. Although the specifier was interviewed immediately after the project finished, he was unable to recount all of his actions, providing a rather generalized account of events, failing to describe the dead ends and being unable to remember how many attempts were made to change the product; behaviour consistent with Yeomans' (1982) findings. While this helped to justify the ethnographic approach adopted, it meant that the interview had to be adjusted to gather the specifier's attitudes towards product selection rather than as a tool to expand upon his observed behaviour.

Although the specifier described himself as creative and always looking out for new products, he was aware that his actual behaviour was contrary to this. He claimed that he was 'forced to be conservative' about product selection and detailing because of his, and his organization's, concerns about building product failure. Products that were new to the office carried a perceived enhancement of risk. His risk management technique relied on the specification of products that he had used previously or, failing that, those used by the office. His collection of literature had been assembled over a long period in the building industry (twenty five years) from products that he said were 'known to perform', i.e. he was pretty confident that, if detailed and implemented correctly, these products would not fail.

Information about a building product innovation would be added to his palette if a new situation had required its consideration and if there were no problems in specifying it and no problems reported from site during or after construction. Trade representatives were only seen or spoken to by telephone when further information was required for a specific project. Thus, communication with building product manufacturers was always initiated by the specifier, which confirms the view that specifiers have clear active and passive modes.

He also said that he tried to stick to products that he had used previously because the time pressures imposed on him by both the design programme and the construction programme rarely allowed him any time to investigate alternatives. At the time of the observation, the specifier was working on three other jobs, all at different stages, and all with demanding programmes, again showing the effect of time constraints noted in Chapter 9.

From this interview, two reasons were noted for the specifier to look outside his palette: technical substitution and a novel design problem. Both of these situations would result in the specifier engaging in an active search for information. Technical substitution would occur if the product in the specifier's palette was unsuitable for the given situation. This would result in a search for information about other products, which may themselves eventually enter the palette. A novel design problem might, for example, occur if the specifier was engaged on a different building type from that normally commissioned, resulting in the need to search for different types of building products.

There was no opportunity for the specifier to use this product again during the period when his actions were being monitored, so the confirmation stage could not be observed. However, when asked whether he would use the product again, he replied, 'Yes, if I need to'.

Reflection and discussion

Before any conclusions can be drawn, it is necessary to comment on the observation method used. Because the author was responsible for the day to day management of the design office in which this individual worked, it is possible that the observer influenced the specifier's behaviour. However, throughout the observation period, the specifier did not seek any clarification of his decision-making process from the author; rather, he sought approval from the technical partner, in accordance with office policy. There is also the possibility that the observer missed part of the process, but retrospective analysis of the written evidence produced by the specifier, namely the drawings, written specification and notes in his desk diary, supported the observations.

This specifier, and other specifiers in this office, preferred to specify products by proprietary (brand) name, and pressures to change associated with this method of specifying have been noted. In organizations where performance specifying is used, the final choice of proprietary product rests with the contractor and, thus, the process and pressures to change will be different from those reported here. Clearly, specifiers working in other design offices will do things slightly differently, and there is a need for further naturalistic forms of enquiry to compare with these findings.

Case study 2: Specifiers B and C

In the first case study the specifier was forced to look for new products because of a detailing problem. Changes to regulations may also cause a change in behaviour, although according to Gann et al. (1998) there has been little research into how regulations encourage or discourage innovation. The introduction of more stringent 'U' values (Part L of the Approved Documents for England and Wales) meant that designers had no option but to change their details and specifications in order to comply with the new regulations; and this provided an ideal situation to observe how specifiers reacted. In this case study (based on Emmitt and Heaton, 2003) two specifiers were observed while independently detailing a cavity wall section for different projects. Both projects were commercial buildings with a very similar design for the external walls. This comprised a steel frame with brick outer leaf, an insulated cavity, and blockwork to the inner leaf of the wall. The specifiers were the first in the office to detail buildings that had to comply with the new legislation.

Specifier B

Specifier B was observed taking the written specification and details from the project he had worked on previously and using the information as a basis for his new specification and details, i.e. he was rolling the project specification. There was no evidence that the specifier accessed the office master specification or standard details, contrary to quality assurance (QA) procedures. However, because the requirements relating to thermal insulation had become more stringent since the previous project was detailed the details could not simply be copied, and so the specifier telephoned the manufacturer of the insulation product he had previously specified (product X). The manufacturer's technical representative recommended an increase in the thickness of their insulating material to resolve the problem. The specifier then adjusted the written specification for the cavity-wall detail by increasing the specified thickness of product X. No attempt was made to alter the standard details, despite the fact that they were now incorrect and contradicted the information in the written project specification. So product X had been retained, albeit in a different thickness to that used previously.

Specifier B issued the details to the structural engineer by fax. On receipt of the information the structural engineer telephoned the specifier and highlighted the fact that the cavity wall ties also needed to be increased in size, together with the foundation details, to accommodate the thicker wall section. (The change also affected the wall to roof section, wall to structural column details and details of all openings, but there appeared to be no discussion about these details at the time.) This helps to emphasize the fact that a small change to one product can have major implications for other building components, making it difficult to consider one product in isolation.

After a short discussion with the structural engineer the specification was revised again to include the larger wall ties (sourced from the usual manufacturer, but a product innovation to the specifier). The specifier made no attempt to check the details that had been provided to him, nor did he make any attempt to consider any cost implications of his decision (the thicker wall required additional and more expensive materials compared with the original detail). There was no evidence that the specifier discussed the issue with fellow colleagues, nor was there evidence of his seeking approval from the design manager, contrary to office QA procedures. Similar to the specifier observed in the first case study, this specifier was also engaged on several other projects, all at different stages, and was under pressure to complete his work to tight deadlines.

At this juncture a second specifier was confronted with the same problem.

Specifier C

This specifier initially took a similar approach to specifier B, rolling the details and project specification from his previous project. Realizing that the details were no longer adequate he also rang the manufacturer of product X. Until this point he was unaware that specifier B had tried to resolve this issue, only realizing the fact during the telephone conversation with the manufacturer's technical representative. On receiving the same advice as that given to his colleague he then had a short informal discussion with his colleague, during which specifier C voiced his concern about the knock on effect of increasing the thickness of the insulation material. They agreed that there was a problem and it was agreed between them that specifier C would investigate the issue further. The specifier then went to the office library and searched the trade literature. Unhappy with what he found (it was all out of date), he telephoned three of the manufacturers listed in the Barbour Compendium that produced thermal insulation products.

The fact that the library information was out of date requires some comment. A reason for forbidding the holding of personal collections of literature is to prevent specifiers from accessing obsolete material. Clearly, it is an essential corollary that the office library material should be properly maintained.

The designer then telephoned each manufacturer's technical department in turn and found that all three manufacturers claimed to have a product that could meet the demands of the new regulations without having to alter the thickness of the insulation (thus the original wall details could be retained). This took approximately twenty minutes of his time. He had not used any of these manufacturers' products on previous projects and so all three represented building product innovations to him.

Each manufacturer sent information to the design office (two by e-mail, one by fax), which contained drawings of typical details, technical details and a proprietary specification for the product. After reading the information specifier C held a brief conversation with the design manager about which product to choose (in line with the office QA procedures). The design manager suggested that cost information be requested from each manufacturer so that a comparative analysis could be made. Information was requested but it was not immediately forthcoming. Instead, all three manufacturers offered to send a technical representative to the office to discuss cost and technical issues with the specifier, but their offers were declined because of time restrictions. The specifier telephoned a cost consultant (QS) and asked for advice on the manufacturers' products. The QS claimed to have knowledge of all three manufacturers, but suggested that one manufacturer be used because they were usually the cheapest of the three. The specifier relayed this information to the design manager, who approved the product, and hence a building product innovation (product Y) was specified. (There was no attempt to compare the performance criteria of these products.) Specifier B then changed his written project specification so that it also included product Y.

It took specifier C approximately sixty minutes to resolve this detailing problem. None of the action observed was recorded anywhere in office documents, other than confirmation of the specified product within the written project specification. A few days later the design manager updated the office master specification so that on all commercial projects the product would be specified (assuming that the specifiers would use the office master files, which appeared not to be the case based on this observation). In this example, the change in regulations did result in the specifiers and the office adopting a building product innovation.

Interviews

Separate interviews were held with each of the specifiers immediately after the observations. The design manager declined to be interviewed. The specifiers claimed that they always worked within tight time frames and so there was little time to assess products adequately, which is why they tried not to change products or manufacturers from those used on previous projects. Both recognized that there needed to be more time spent on the office details and specification, but that they did not have the time to do so. They recognized the danger of rolling specifications and details from one project to another and releasing incomplete (and contradictory) information, but claimed that 'everyone did it' out of necessity, not through choice. Both were critical of the way projects were programmed within the design office, claiming that with every project the demands to produce information became ever more demanding and the time allocated to complete the task was inadequate. Both specifiers, however, recognized that this was not just a matter of better programming, but that time was short because of the downward pressure on professional fees; thus, some responsibility had to rest with clients.

The habit in this office was also to use proprietary specifications. Both specifiers claimed that performance specifications were only useful if written very tightly, thus limiting the contractor's choice to one or two options. Both specifiers were uneasy about passing the choice of product down the chain to the contractor, their perception being that the quality of the building would suffer. They also said that they would not use performance specifications for important details, such as cavity wall insulation, because they were concerned about liability if the product failed. Given the rather hurried manner in which the insulation product was selected in these observations it would not be unreasonable to conclude that a performance specification might have been a better option.

It was not possible to observe what happened to product Y during the subsequent stages of the project. However, it was possible to revisit the office on completion of the project and speak to specifier C. He said that product Y had been built into the building, without any problems or as far as he could remember any attempts to change the product. However, he said that shortly after the projects had started on site the original manufacturer of product X had contacted the office to say that they had revised their product range and they now had a similar product to their competitors. Specifier C said that is it was highly likely that the office would revert to their original manufacturer and use the new product (product Z), indicating office loyalty to this particular manufacturer.

Reflection and discussion

The data reveal an insight into specifiers' behaviour under time pressures and the failure to use office management procedures. Decisions were made within very tight time frames while specifiers were engaged on other projects. There was very little time to consider and reflect on the consequences of their actions. Both specifiers were under considerable pressure to produce information quickly and both took slightly different approaches to product selection. Neither specifier made any attempt to check the technical information provided to them by the respective manufacturers or the cost advice given to them by the QS. Quality management procedures were not followed; neither was there any attempt by the design manager to emphasize the importance of standard procedures. Instead, the design manager appeared to operate in a 'fire fighting' capacity, answering questions from the designers when necessary. It was evident in the discussion with the specifiers that they were not particularly happy with the manner in which the design office was managed. There was no evidence of any formal knowledge sharing practices; instead, designers relied on infrequent informal conversations with colleagues in the office and telephone calls to actors in other offices. Unfortunately, because the design manger declined an interview, it was not possible to obtain his perspective on the specifiers' actions.

Case study 3: Specifiers D and E

In this design office data were collected over a six month period using participant observation (based on Emmitt and Johnson, 2004). The habit in the office was to roll specifications from one project to the next, despite the existence of an office master specification and quality management procedures. In an attempt to observe the specification of building product innovations the researcher identified four projects as potential case studies because they were at the detailing stage. A specific element was selected from each project, being the detailing of (1) windows, (2) pitched roof, (3) ground floor/wall junction, and (4) entrance porch detail. In taking this course of action the designers working on these particular projects became the focus, and thus they were self selecting. This limited the observations to two specifiers, specifier D and specifier E. Following the observations both specifiers and the design manager were interviewed.

Specifier D

The first example was taken from a design for a single storey residential building complex. The mechanical and electrical (M&E) consultants were heavily involved in this project and proved to be a major source of information relating to the performance requirements of the windows. The M&E consultant telephoned the specifier to inform him that the chosen window specification (aluminium framed) did not comply with the U value for windows with metal frames. The office had specified this window on previous commercial projects without any questions being raised, and so there was some uncertainty about the product's performance. In an attempt to resolve this, the specifier telephoned the manufacturer (manufacturer A), who claimed that their windows did comply and that they were working on calculations to prove it. Later that day a second manufacturer (manufacturer B) telephoned the office in an attempt to get their windows specified. The sales representative had heard from an 'external source' that the office had problems with their usual manufacturer. (The external source was never revealed, although it was probably the M&E consultant who contacted manufacturer B.) The technical representative was invited into the office to discuss the technical requirements of the alternative window system. He delivered to the office a sample of the product and full technical details, but no cost details. Following a short meeting, the manufacturer was given verbal acceptance by specifier D and left the office with a copy of the design drawings. Manufacturer B's proprietary specification and detailed drawings of the window arrived a couple of days later and were incorporated into the written project specification by specifier D.

This helps to demonstrate how quickly a change can be made and how informal communication networks facilitate the decision making process and the subsequent adoption of a building product innovation. Manufacturer B was unknown to the office and has since become one of a number of preferred suppliers. Manufacturer A eventually came up with the calculations (two weeks after they were required) and, although rejected on this project, their window system was specified on later projects by this specifier. So although he had been forced to change his preferred product on one project he reverted to his original choice once he had regained confidence in it. There was no evidence in the observations that costs were discussed. The specifier did not discuss the change with the design manager; again, this was contrary to the office QA procedures.

The second case related to the detailing of a pitched roof. The problem here was that a sloping ceiling detail did not allow enough space for the required depth of thermal insulation. Specifier D initially discussed the problem with a colleague. He was concerned about constructability and aesthetics, as well as complying with the regulations. His colleague claimed that he was aware of a new insulation product that may solve the problem; he could not remember details, but knew that a contractor known to the office had used the product. He gave the name of the contractor to specifier D, who then telephoned the contractor for details, which were forthcoming. Specifier D then contacted the manufacturer directly to obtain a sample and technical details. The information arrived the following day, was discussed briefly with the design manager, and the product specified on a trial basis (in conjunction with products already known to the office). Both the specifier and the design manager had reservations about the perceived high cost of the product, although at no time in the observations had costs been provided by the manufacturer.

Specifier E

A wall to ground junction had to be detailed by specifier E for two commercial projects with similar types of construction. The intention of the design manager was to use this detail as an office standard once it had been drawn, and therefore there was increased attention to this detail. Specifier E copied a detail from a manufacturer's technical literature (newly received in the office), but altered the detail so that the insulation continued in the cavity below the damp-proof course. This appeared to compromise the manufacturer's warranty and the design manager questioned the detail, especially the wisdom of carrying the insulation below ground level in the cavity. To obtain technical advice specifier E telephoned the manufacturer to seek reassurance about the performance of the product below ground. This was forthcoming and the detail was issued as an office standard to be used on the current and subsequent projects. The decision to use the product was based entirely on verbal information given by the manufacturer's technical department, which seems rather a risky strategy. There was no attempt to explore alternative manufacturers.

The fourth case study involved the detailing of an entrance to an educational building. Two designers in the office could not agree how best to detail the canopy to comply with the Building Regulations, and therefore specifier E telephoned the building control officer for advice. The building control officer gave a vague answer, claiming that it was not his job to advise the office. The specifier then telephoned a number of manufacturers known to the office to request technical details of their products. These were new products from familiar manufacturers and a rapid decision was taken to specify one of these. There was evidence of seeking advice from fellow designers and the design manager, as well as from outside the office. None of the designers consulted the approved documents throughout the observation period, relying solely on information provided by the manufacturer.

Interviews

Both specifiers were aware of their reliance on manufacturers' information and informal guidance and to a lesser extent on building control officers.

They claimed that they 'had to trust' the manufacturers, given the lack of time to investigate products thoroughly. The specifiers relied heavily on the manufacturers producing the technical details and written prescriptive specification for brand products. This was primarily to save time, but they also recognized that the manufacturers knew more about their products than they could hope to within a very limited time frame. Manufacturers who were willing to produce the detailed drawings and the written specification 'for free' were specified in preference to others. Once specified, they were reluctant to change their selection unless forced to do so. This supports the earlier research into specification decisions and emphasizes designers' preference for proprietary specifications for certain building elements. Relationships with manufacturers were informal, and may be described as a form of informal alliance.

Both specifiers raised (unprompted) the problem of managerial control in the office. Responsibility for the quality of the production information and the management of the production information was left to the discretion of individual designers. Specifiers said that management issues were discussed at the monthly staff meeting, but that time and resources were always cited as reasons to not implement improvements in design management processes. Other than informal communication between members of the office, there was no means of sharing knowledge between individual members of the office. Indeed, the specifiers both claimed that they had little knowledge of what their colleagues were doing, unless they were consulted to discuss a problem.

The design manger agreed to be interviewed. He was aware that procedures implemented to help specifiers were 'sometimes followed', depending on time pressures. He claimed that he had to adopt a flexible approach; otherwise tasks would not be completed within the time frame. He claimed that professional fees were declining in relation to the amount of work required and that this placed unreasonable pressures on the architects and technologists within the office. Like the specifiers, he was aware that the office used practices that were viewed as unprofessional because of the tendency to promote errors from one project to the next. However, he was quick to point out that the majority of design offices had to operate like this and that his peers in other offices were no different in this regard (a claim supported in the small amount of published research). Furthermore, the office had a good reputation for delivering good quality buildings on time and within budget, so it was not necessarily a problem.

Reflection and discussion

That designers relied heavily on information provided by manufacturers, often picking up the telephone to ask questions and to seek reassurance rather than reading through technical literature, supports earlier observations of Mackinder (1980) and Emmitt (1997). There was no evidence of any designer checking the information provided by the manufacturers. Furthermore, there was no evidence that the designers consulted the approved documents or relevant codes, although they were available in the office, relying solely on information from others. Specifiers were primarily concerned with constructability and aesthetics. Costs were discussed but were not given high priority. Specifiers were responsible for the management of their work, with no evidence of direct supervision from the design manager; nor was there any evidence of managerial processes that would help with the decision making. Only when a problem occurred did the specifiers consult the design manager.

The lack of checks on manufacturers' information perhaps warrants some comment because this seems to be a risky strategy. It raises questions of liability in the event of a failure, especially given the interconnectedness between building components. And it seems to be a curious pattern of behaviour given that the use of familiar products is seen as a way of reducing risk. One must suppose that designers fail to check manufacturers' claims because of the time pressures when specifying and that their common experience is that the information provided is generally correct. So it might be in the majority of cases, but the experience of the failures of trussed rafter roofs discussed earlier shows that it is possible for manufacturers to get things spectacularly wrong.

Discussion

All of the specifiers were working on other projects, all at different stages, and all with demanding programmes, and therefore the potential for investigating manufacturers' claims as to the performance of their products was very limited, serving to reinforce the established products. Again, there are parallels with medical research. Studies into repeat prescribing (Harris and Dajda, 1996) found that medical drugs were prescribed, without further reference to the doctor by the patient (primarily to save time), thus reinforcing the use of a familiar drug. Like the patient's drugs, the products have not been reassessed, merely applied because they worked successfully before.

One of the issues highlighted through the observations was the impact of other parties on the specification process, a characteristic present neither in the studies of repeat drug prescribing nor in the large body of diffusion of innovations literature. At different stages in the innovation decision process, contributions were made from outside the architect's office by individuals with different priorities from those of the specifier. Because pressure to change specifications is something that a specifier has to deal with, not just during the design phase but during the assembly process as well, there appears to be a need to add two sub stages to the Rogers model (Fig. 7.2), to accommodate the uniqueness of the specification process. Adding one between stages 3 and 4, stage 3a would cover the pressure to change products during the tender stage; adding the other between stages 4 and 5, stage 4a would cover the contractor's attempts to change products. This is discussed further in Chapter 11.

Reflection on the method used

The actions reported above were influenced by the specifier's office environment, time pressures, characteristics of the project and the characteristics of the specifier, and therefore it is necessary to consider how representative this behaviour is likely to be of other specifiers. When talking about their behaviour, their opinions were consistent with Mackinder's sample of architects, despite the long time gap between the two pieces of work. Moreover, as their behaviour was also consistent with that of other specifiers reported in Chapter 9, there was no evidence to suggest that the actions recorded were unrepresentative.

The observer was actively involved in the social environment in which the data were collected and thus may have affected the process in some way. It is also probable that the observer missed events vital to the processes being studied, partly because of the difficulty of seeing and hearing everything that was occurring in the office environment and partly because there were times when the observer was out of the office, engaged in other tasks such as visits to construction sites. Final decisions were recorded in the written specification and encoded in the working drawings and architectural details; there was no evidence of any of the preceding discussions and actions observed during the monitoring period in any of the written documentation. The interviews were designed to identify any gaps in the data collection and analysis of documents produced by the designers, e.g. drawings, written specifications and notes in desk diaries, which helped to confirm that the data collected were as complete as could be expected given the limitations of the method used. When interviewed, the specifiers failed to recount the process as it happened. Events were recounted as a simplified version of that observed, or had been forgotten. When prompted by the interviewer, the designer simply said they had no recollection of a particular activity. In some respects this justifies the use of the participant observation method and highlights the fact that detailing many different parts of a building is not something that readily sticks in the memory. Thus, asking questions about the process after the event may provide misleading data. While the findings of the observations are consistent with previously published work into specifiers' behaviour by Mackinder and Emmitt, it is important to recognize that this observation method can only identify issues specific to the situation observed; other specifiers may act differently.

Design management issues

The design offices in which the observations were conducted were well known design practices, employing qualified staff. All five specifiers were experienced professionals, male, and aged between thirty and fifty, and had similar educational backgrounds. Although there were differences in age and experience this was not reflected in their actions. In case study 2 one of the specifiers acted in a more considered manner compared with his colleague. The offices had a reputation for producing good quality architecture and for delivering projects on time and within budget. All of the offices had quality management systems in place, but these tended to be ignored by the specifiers because they perceived the procedures to be too time consuming. Individuals were working to project specific deadlines and were left very much to their own devices as to how they met the programme milestones. All of the specifiers expressed a desire for more time to be allocated to the detailing and specification stage. The use of management systems and day to day managerial control within the offices tended to be inconsistent and it was evident that the specifiers wanted better guidance from their design managers. Design managers were only consulted when the specifier faced a problem that could not be resolved in isolation. Thus, the design manager had a relatively minor influence on the majority of specification decisions.

Specifiers claimed that professional fees were too low for the amount of work required to produce satisfactory information. The result was very limited time for the completion of working drawings and associated specifications. This made it difficult for designers to spend adequate time assessing new products and brought about increased reliance on manufacturers' information. The lack of time was also blamed for rolling specifications from one project to another.

Specifiers preferred to use proprietary specifications, relying heavily on manufacturers' brand specifications and details. The specifiers and the design manager felt that performance specifications were more suited to large, repetitive projects and were too time consuming to use on small to medium sized projects. They were highly sceptical about using performance specifications and passing the responsibility for product selection to contractors (who they appeared to trust less than the manufacturers). Their apprehension related to concerns for the quality of the building and a desire to protect the interests of their clients. There was no evidence in these observations that specialist sub contractors were consulted over specification decisions, and there was only one example of a contractor being consulted during the detailing phases. In these observations influence on the choice of product came from members of the design team (e.g. structural engineer and QS) and the manufacturers.

Knowledge sharing within the design offices was not carried out in a strategic manner. Specifiers were engaged in individual tasks and would only communicate with others if they faced a problem. From a researcher's perspective there appeared to be a need for the specifiers to talk about their good and bad specification practices on a regular basis, perhaps through a regularly held knowledge-exchange meeting. Approaches to specification were highly individual and at times rather inconsistent. Although the design managers operated a relatively relaxed approach to the office management systems to enable the office to function, it is questionable as to how effective the management systems are. Again, from the researcher's perspective there appears to be a need to redesign the management practices to create a better fit between the way in which designers operate and the requirements of an effectively managed office.

The partners and design managers were, at the outset of the research, somewhat perplexed that anyone would want to research the specification process. They saw it as a very familiar aspect of the designer's job and hence unworthy of research effort. When presented with the research findings the design managers were surprised by the results. One design manager refused to accept that 'his' designers were operating in the manner observed, and subsequently refused an interview. Similarly, one of the other design managers expressed some initial surprise over the findings, but claimed that it was 'understandable' given the very limited time available to detail the building. When presenting and discussing the findings with the partners they appeared to be more pragmatic. They claimed that there was a difference between getting the work done in the time available and doing it as prescribed in the textbooks. Despite complaining about their fee income they were quick to point out that they still performed well and that they were not constantly making errors. This is confirmed in the observations. While it would be reasonable to conclude that the specifiers were not always following good practice, their relatively informal working methods provided a culture in which other actors were quick to point out potential errors. This informal communication network appeared to be based on mutual respect and trust. All of the projects observed proceeded to successful conclusions (completed on time and within budget).

Implications

The observations reported above help to illustrate some of the pressures and the complexity of the decision making process through which specifiers pass. They help to illustrate communication networks and pressures on the design process that were not evident in earlier research. The findings also suggest that building-product selection is a very personal issue for designers and, as noted above, a difficult process to observe.

Although the research reported here is limited in its scope, it has helped to shed some light on the complexities of the detail design decision making process. The value of the close observation, possible with a participant observer, is that it allows us to understand better the detailed behaviour of specifiers in the design office, which is difficult, if not impossible, with other methods. The research identified a difference between the observed behaviour and the perceived behaviour of the specifiers. The findings also help to illustrate the complex and informal relationships and decision making that go on within the design office. The value to the design management field lies in the detail, which supports the more general literature that calls for better management of creative offices. More research into the everyday working practices of designers would be useful in helping to identify issues, from which design managers could apply simple and pragmatic process improvements.

Rogers (2003) questioned whether it was the need for an innovation, or the awareness of an innovation, that comes first in the innovation decision process. The research reported here suggests that specifiers actively search out building-product innovations only when the need arises and not before. The implication is that the adoption of 'new' products may face considerable resistance, not just from the specifier but also from the other contributors to the specification process. By gaining a fuller understanding of the individual's innovation decision process, professional design offices may be in a better position to manage this critical aspect of building design. To do so, however, requires further research, both to test the results presented here and to further our understanding of this little analysed aspect of design decision making. Instead, what we can do is propose a model of the specifier's innovation decision making process based on the research reported here. This is described in Chapter 11.

Towards best practice

This final chapter brings together theoretical and practical considerations for all those concerned with specifying buildings. As demonstrated, factors both internal and external to the specifier's social system may exert different pressures at different stages during the project. Thus, although specifiers are mainly content to choose from their preferred palette, building product innovations may be forced on them by pressure from, for example, planning officers or contractors. Therefore, building product innovations may be introduced by sources other than the manufacturer.

At the outset of this research, it was assumed that the adoption of building product innovations was, to a large extent, influenced by the communication of information from manufacturers to specifiers, but it has become apparent that this process is more complex than that. Although communication of information about building product innovations from the manufacturers is continuous, specifiers use selective exposure. That is, they operate personal gatekeeping systems where a gate is only opened when they are actively involved in product selection. In addition to this personal gatekeeping, it has been shown, albeit from a small sample, that the partner of the architect's office exerts a considerable influence on the amount of information allowed through the office's gate to the specifiers in the office. Potential specifiers may not be aware of building product innovations, although information on them has been sent to the office, simply because they failed to pass through the formally established gates. Therefore, they cannot be considered for adoption unless knowledge about them is gained from another source.

However, even this model is too simple, because it has been demonstrated earlier that specifiers continue to select from a palette of favourite products unless forced to look for alternatives when familiar products cannot meet some particular requirements. This suggests that the individual's informal gate is closed for a large proportion of the time, and only opened in specific circumstances. It is these specific circumstances that are of interest to manufacturers and design managers alike.

As argued above, a set of conditions seems to be needed before the innovation-decision process can begin. If the palette is adequate, then the specifier will not have to spend time searching for alternatives. If, however, the palette is inadequate, i.e. cannot solve a particular problem, then the specifier will be forced to search for unfamiliar products, thus starting the innovation-decision process. It also appears from Rogers' model that this process can begin if and when the potential adopter becomes aware of the innovation passively. However, given the time pressures exerted on a specifier, it is unlikely that he or she will investigate information simply out of curiosity.

A model of the specifier's decision making process

The model presented here is based on Rogers' model and helps to illustrate the complexity of the process, but it was found necessary to modify this to take account of time pressures on the adopter in this process. Knowledge of this should help design managers to manage better the specification process through greater awareness of the issues. Furthermore, it should help building product manufacturers to reconsider their marketing strategies to specifiers (Fig. 11.1).

Prior conditions

When selecting products, it seems that specifiers have (1) existing personal experience of products used previously, represented by the palette of favourite products, and (2) knowledge that information about other products, building product innovations, may exist in trade journals, product compendia or the office library. In the case studies it was always the failure of the specifier's palette of favourite products that led to a search for information on unfamiliar products to satisfy a particular need. So, although information about products had previously been sent to the office, via journal advertisements and listings in the product compendia, manufacturers must depend on the particular circumstances of the design and specification process for specifiers to become aware of their products. Specifiers must be actively searching for innovations or be made aware of them through third parties. In building design, specifiers are not passive adopters; rather, they search for innovations only when the need arises. This occurred in the following stages.

Stage 1: Knowledge

In most cases knowledge was first sought from colleagues in the office before looking in the library, to save time. This requires some discussion in terms of the Rogers model because it is the office that is the unit of adoption. The file

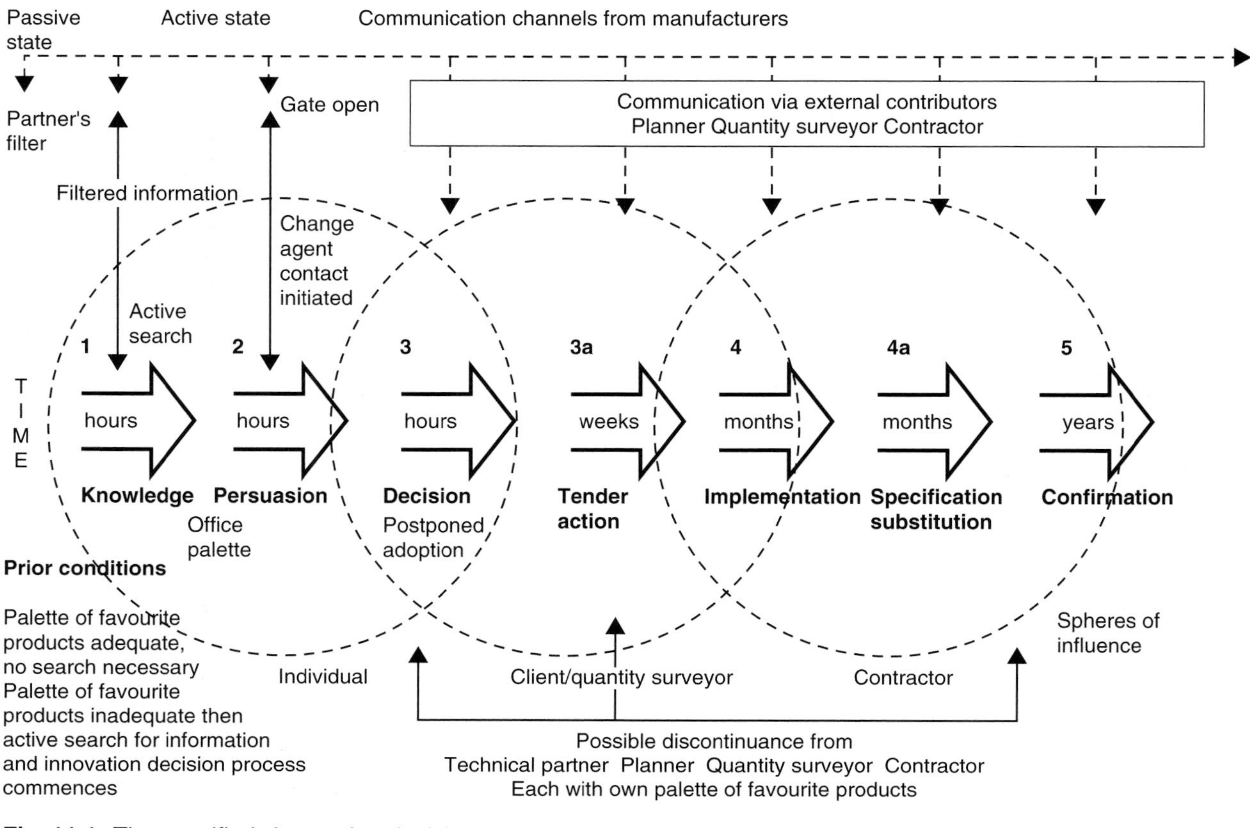

Fig. 11.1 *The specifier's innovation decision process and spheres of influence*

of filtered and therefore approved products is available; these products are seen as carrying less risk than those rejected in the filtering process. The office has become aware of these products, even if the individual specifier has not. There are also products that have been used previously by those in the office, which will largely be a subset of the above. When information is sought from colleagues, the tendency will be for the specifier to investigate previously used products first, so that adoption of these for another project will constitute confirmation of adoption by the office.

Only if this product is not suitable in the particular circumstances will the specifier search the office library, leading to the consideration and potential adoption of an innovation. There are actually two kinds of innovation here. There are those that are innovations to a particular specifier, but not to the office as a whole, and those that are innovations to the office. The office can be said to have an awareness of them because information is within the office library and has passed through the hands of the technical partner, but they have not yet been adopted because the circumstances favouring their adoption have not yet occurred.

Stage 2: Persuasion

Specifiers 'tested' the companies' technical departments, making initial decisions based on a telephone conversation, before deciding whether or not to continue investigating the product and also whether or not to invite the trade representative into the office. This decision relied largely on the telephone conversations with the technical department during the persuasion stage. Testing each product against the particular situation may involve the making of sketch details to see whether the product is suitable. This may be regarded as a trial of the product, although not in the form that Rogers envisages, i.e. it is a trial in theory rather than in practice. Nevertheless, this is an important feature of the design process. At this stage, the specifier may be only partly satisfied with the product or be in a state of uncertainty because of insufficient information, and therefore may invite the manufacturer's representative into the office to help and explain. Thus, the persuasion takes place in more than one stage.

Stage 3: Decision

The decision to adopt the building product innovation was influenced by other parties and was far more complex than any of the examples reported by Rogers. For example, the influence of the quantity surveyor (QS) was influential in bringing about a search for a cheaper alternative and awareness of further building product innovations. The greater complexity of the process is a direct result of the greater complexity of the social structures compared with those examined by Rogers, and this was also seen in subsequent stages. It should also be noted that if this product has already been used by the office (by a different specifier), then adoption is simply confirmation of its earlier adoption.

Stage 3a: Tender action

Further building product innovations were introduced to the specifier by the QS and the contractor after the initial decision to adopt the innovation had been made. This stage needs to be introduced because of the competitive tendering (bidding) aspect of construction projects. There were several opportunities at this stage for both the discontinuance of an innovation and the adoption of alternative products. Since the options introduced by the QS and the contractor were also viewed as product innovations by the specifier, this shows that building products can be marketed successfully to different members of the building team.

The case studies confirmed the existence of this additional stage, but the adoption of an alternative product cannot necessarily be considered as discontinuance of the innovation by the office if the product originally specified is now rejected for this particular design. Discontinuance by the office would only occur if the product now introduced were to be adopted as a permanent substitute, i.e. to be used in all subsequent designs. An aspect of building product innovations different from the adoption of other innovations is that it does not imply total rejection of all other products. While a farmer might only sow one type of seed, thus making a clear choice, building products always remain as one option among many, each used in appropriate circumstances.

Stage 4: Implementation

Inclusion of a building product in the documentation from which the contractor was to build the building, the written specification and drawings, is, from the specifier's perspective, implementation of the decision to adopt. However, this was not the end of the process because contractors made repeated attempts to change products during the construction phase, thus confirming the requirement for a further stage in the innovation decision process, stage 4a, when considering the adoption of building product innovations.

Other professional activities may also involve decisions taking place over time and when more than one product is involved. A doctor may prescribe a new drug for one patient but may remain with a tried and tested product for another. During a course of treatment, that doctor may also switch from one drug to another for a particular patient. It is clear that the adoption process by professionals who are making decisions for a range of clients in different situations cannot be expected to show consistency of adoption. One might doubt the flexibility and thus the competence of a professional who rigidly adhered to a single product, and thus a more complex model is required for professional adoption processes.

Stage 4a: Specification substitution

Despite the use of a gatekeeping system, once the project was in the contract stage, it was possible for the contractor to introduce further building product innovations to the specifier, bypassing the gatekeepers because they communicated directly with the specifier. At this stage, the newly introduced product was assessed against that already specified, but was resisted because of the time required to investigate its characteristics before use. However, the possibility of substitution at this stage clearly exists and is known to occur from anecdotal evidence. Therefore, adoption is only completed when the builder actually uses the product on site.

Stage 5: Confirmation

It has already been noted that the adoption of products previously used by the office, but new to the particular specifier, constitutes confirmation by the office. In the research reported earlier confirmation was not observed, although the specifier placed the information in his palette for future use. Confirmation by the office might also occur through the subsequent use of the product by another specifier within the office.

The case studies were mainly concerned with product innovations that were to be used on the exterior of the building, and so would be visible when the building was completed, which perhaps explains part of the specifier's reluctance to change his original decision. However, the specifiers said that they rarely had time to investigate 'new products' because of the pressures of the building programme, and noted that this aspect of the job, while important, had to be balanced against many other duties. This behaviour was consistent with Mackinder's sample and also with the respondents of the postal questionnaire, who tended to act conservatively and rely on products used previously, both to save time and to reduce risk.

Spheres of influence

The contribution of external agents to the innovation decision process was far greater than had been anticipated. The main contributors were the client, consultants (the QS, mechanical and electrical consultant, and structural engineer), the planning officer and the contractor. Influence was exerted at different times during the process, but led to both the discontinuance of adoption and the introduction of knowledge about further building product innovations. These two kinds of influence are clearly linked, since one may lead to the other. This kind of pressure comes from those with quite different values and priorities from those of the designer.

Awareness of product innovations from contributors other than building product manufacturers can occur at any time after initial specification, reinforcing the contention that the process is more complex than the Rogers model. These external contributors' motives for change were sometimes driven by cost factors, a characteristic of building products that is largely unconsidered by specifiers until they are forced to do so by others.

The degree of influence from outside the specifiers' office has implications for offices because the formally established gates are bypassed. Add in the pressure to change decisions quickly, and even the best managerial systems can become ineffective control mechanisms. Clearly, for the design manager, it is important constantly to remind specifiers about the importance of their specification decisions and the likely consequences of changes suggested by parties from outside the specifiers' office.

Research matters

Before drawing some conclusions from this research, it is necessary to consider briefly the limitations inherent in the work and thus the limitations of those conclusions. The first point to note is that this research has only been dealing with buildings designed by architectural practices, although there are important sectors of the building industry in which they are seldom involved. This includes contractor led contracts with their in house design departments and the housing sector, where the kinds of professional office described here are not involved. Housing is dominated by a few large builders, but with a very large number of small builders, the latter being notoriously conservative. Bowley (1960), for example, has described how difficult it was for the plasterboard manufacturers to persuade builders to use their product. At first, they considered plasterboard inferior to the use of wet plaster and also failed to understand how it should be used. However, its eventual success depended on its uptake by the house building sector of the industry, and Bowley regards its widespread adoption in the 1930s as an example of prefabrication in the industry. In contrast, although a number of novel methods for prefabricated housing were introduced after the First World War, to solve the serious under capacity of the traditional industry, this industry eventually reasserted itself, and most of the prefabricated systems were abandoned.

Nevertheless, within this conservative sector, dominated by the use of brick and block walling and timber upper floors and roofs, a number of innovations has taken hold. These have been the eventual adoption of plasterboard in the 1930s, the use of slab on grade ground floors replacing suspended ground floors in the late 1950s and, more recently, the fairly widespread adoption of precast concrete flooring. Additional innovations have been the use of the Timber Development Association (TDA) roof and then trussed rafter roofs in much the same period and, to some extent, the adoption of timber frame construction in the 1970s. The adoption of the TDA roof and interlocking roof tiles at much the same time was in response to timber shortages following the Second World War (Yeomans, 1997). The return of suspended ground floors, although not necessarily of precast concrete, was brought about because of the shortage of flat sites in more recent years. The impression given by these few examples is that the adoption of innovations by builders is largely in response to particular external pressures, but there is no real evidence to substantiate this as no detailed observation of builders has been made.

One of the effects of the recent trend towards increasing prefabrication is not only to move production from the site to the factory, but also to move specification from the individual design office to that of the manufacturer. Both the reasons for and the effect of this trend need some exploration that lies outside the scope of the work carried out here. If units, such as housing, can be standardized then the reduction of site work, with greater rapidity of building and easier quality control, has clear advantages for the speculative builder.

As contractors were having an influence on the process observed, they were clearly becoming aware of new products, presumably through having to use them in response to their specification by other designers or through their use on design and build contracts. Again, their positive or negative reactions to these new products have not been observed. There were also other professionals who introduced new products to the specifiers, and this research did not reveal the process by which they became aware of them, or what factors affected their responses.

Research methodology

This research has been largely based on the direct observation of specifiers' behaviour, supported by interviews. Very little of the process reported was recorded on paper during the normal course of work in the office, and it is clear from the account given that not only would trying to trace the process after the event have been very difficult, but any results would have been potentially misleading. There is always a problem about trying to trace these processes after the event because those involved are likely to remember a far more abbreviated process than that actually followed. This has already been observed in the design process, when the designers' own impressions of the process after the event were compared with a more objective account based on data collected during it. It was then found that the designer tended to describe a logical and linear process, rather than the iterative process actually followed. Dead ends, which might well have been important as part of a learning process in the design, were simply forgotten (Yeomans, 1982). This confirms Rogers' recommendation that the adoption of innovations should be traced during the adoption process itself and not afterwards.

It had been possible to obtain a record of the design process used in the above experiments because of the rather particular circumstances of a 'team design', where thoughts had to be articulated. Note, also, that the behaviour of one of the specifiers observed and reported in the first case study was made possible because of rather unusual circumstances: his habit of talking to himself. The circumstances peculiar to these observations mean that it is not normally such a simple matter to carry out work of this kind. Researchers who wish to follow up on this kind of study may need to adopt a participant observer approach, essentially the research method adopted by workers such as Cuff (1991).

Here, the methods chosen were quite different. The observations were made over an extended period with no direct manipulation of the process. Because normal duties in the office meant that the observer was not always present, there is the possibility that significant events might have been missed. However, there was insufficient preliminary understanding of the process to enable the design of a more controlled experiment, so this 'natural history' approach appeared to be the most satisfactory. In this case, the fact that the specifiers were being observed was not apparent to them. In all of these cases, no attempt was made to change the behaviour of those being observed, nor was there any obvious temptation (or possibility) for those being observed to seek advice or comment on their actions from the observer. It is, of course, possible to envisage experiments where this is precisely what is done, i.e. where the observer is there not simply to observe, but to exert some influence over the subjects of the observation in order to 'improve' their behaviour in some way.

Action research is one method of deliberately effecting change in a social system (Lewin, 1946) and could be applied to particular parts of the specification process, for example to the product selection or writing stage. Typically, action research involves active participation by the researcher, working with those under study to identify the problem. Once this has been done, solutions to the problem can be suggested, implemented in the workplace and monitored over an agreed period. The data can then be analysed and the solution to the problem evaluated to see whether it has, or has not, brought about improvements. This applied research method relies on the researcher and those under study trusting one another, since the researcher will be involved in bringing about change in a social system in which he or she does not normally work. Action research also relies on the specifier's office recognizing that there is a need to improve working methods and this might be a stumbling block, since many of the offices visited in this study neither recognized the importance of specifying nor saw the need for change.

Establishing and maintaining best practice

There is a very clear need for improving awareness about the importance of specifying buildings correctly, both for individuals and for design and engineering organizations. Clients also have an interest in improved specification practices, since they are paying for them. There are two interrelated areas to consider here, professional updating and reflective practice, key elements to the ideal of life long learning.

The more efficient the specifying process, the better it is carried out, and the more time and resources can be spent on other (more profitable) tasks. Professional updating via continuing professional development (CPD) initiatives and in house training programmes can help to refine and improve the manner in which tasks are carried out. Some of the larger design organizations organize their own in house updating sessions for their staff. Smaller organizations have to rely on informal networking arrangements or on formal CPD sessions organized by a professional institution and/or a commercial enterprise. This might help to rectify the poor treatment of this aspect of design in schools of architecture. Moreover, the recent introduction of the vocational qualification in Architectural Technology at Level 4 (NVQ) may help to improve specification practice because it contains an entire unit dedicated to specification writing and decision making. However, to the authors' knowledge theirs is the only work that has been carried out on specifiers' behaviour in practice.

Linked to continual professional updating is the philosophy of reflective practice; the ability to reflect on one's actions in the workplace with the aim of doing things better the next time. The argument for reflective practice is well rehearsed and its effectiveness clearly demonstrated. However, time must be found to engage in professional updating and reflective activities, and to programme them into the working pattern of the design office. This is something for the design managers and owners of the firms to consider in their management of the business and a number of related issues can be identified.

The need for comparative information

Rogers has characterized the attributes of innovations that make them more or less likely to be adopted, and these have been discussed in relation to building products. However, the comments above suggest that a number of other attributes may be influential at different times or in different circumstances. Consider durability as an example. Under most circumstances, one would assume that this is an important attribute, as buildings tend to last for a long time. Apart from in temporary or relatively short lived buildings, durability is an important attribute of both buildings and their individual components. However, there are also buildings that in themselves are durable but which contain parts that are expected to be replaced at frequent intervals. In shopping complexes, the structure may be 'permanent', while the shop fittings will be frequently changed to keep up with modern fashions. This implies different selection criteria for different products within the same building.

Time was a recurring factor in the observations, with specifiers severely limited by time constraints. This is particularly true of designers working on refurbishment and rehabilitation projects, where the choice of product cannot always be determined until the work is opened up. Here, the ability to make a decision in limited time is vital and tools to help designers to access reliable information quickly, so that they can make informed decisions, would be welcome. However, even with the growth of information technologies and the trend towards electronic gateways, where filtered information is passed on to design organizations on a 'pay to use' basis, there is still no source of comparative data to assist the specifier. Indeed, whether there will ever be such a directory is questionable.

There are a number of reasons for this. First is the sheer enormity of the task; the number of building products available means that there may be too many to compare. Reducing the number of products compared and listed in a directory would make the task easier to achieve, but to be useful such a directory would have to be comprehensive and this raises the problem of keeping it up to date. New products and minor product improvements are continually introduced, and the information would need to be monitored and revised frequently to maintain its currency. While the information can be updated quickly in electronic format, the actual process of listing the products characteristics against established benchmarks would be time consuming. A major obstacle would be the task of checking the claims made by individual manufacturers in their promotional and technical literature. A reliable, independent, organization would be needed if the information were to be trusted by specifiers, which would have to be paid for. Such a directory might help them to narrow alternatives, but they would still need to access, and perhaps communicate with, the manufacturer in order to resolve queries relating to the specific issues of particular jobs.

Different levels of quality

One could be forgiven for thinking that the best product must be selected every time. The research findings reported earlier clearly show that the product's characteristics are only one of a number of issues considered by the specifier. With concerns over quality and liability, the focus is usually on the product perceived to be most durable and least likely to cause the specifier problems in the future. Indeed, with the ever present threat of legal action against design organizations, the natural tendency is to over specify as a means of protection. Thus, buildings may be more expensive to build than need be; indeed, the lack of comparative data discussed above may well exacerbate the situation.

Questions have to be asked about the quality required by the client for a particular project and the quality actually specified. These discussions should form part of the client briefing stage and, once agreed, the desired quality for the building and specific elements of the building should then be confirmed in the project brief. This will form a point of reference for designers and specifiers and is essential if value management exercises are to be conducted effectively. Recently, an attempt has been made to try to establish an information system based on different quality levels. Specialist contractor procurement files (Dixon, 2001) aim to provide generic specifications for a product type based on the collective experience of manufacturers, installers and specifiers, providing three different specifications to suit different quality levels.

Environmentally friendly (green) products

Sustainability has recently become an aspect of the selection process, the concern being that products should be selected that consume less energy in their manufacture or are in some other way less environmentally damaging. Because this is a relatively new criterion, it may be assumed that, while there may be some products that have always been satisfactory in this regard, there will be many that are not so. Thus, if specifiers are behaving in a responsible way, one would expect to see a shift away from some of the older, less sustainable products towards products that satisfy the new criteria. (There may, of course, be improvements within familiar products that make them more sustainable.) This implies a period of greater innovativeness among specifiers as they search for these new products.

There is no evidence that this is happening. Indeed, there is every indication that it may not be so (Peat, 2007). While the adoption of these new products may have advantages for the planet, there is no indication that they have any advantage for either the specifier or the individual client, and the reverse may be true. The adoption of such products may increase the risk of failure, or simply be perceived to do so. Social conscience does not necessarily overcome a tendency to continue to use familiar products, and history suggests that this is even so when financial incentives are applied. At the turn of the nineteenth century it was government policy to encourage the purchase of North American timbers rather than Baltic timbers, which they did by differential import duties. Nevertheless, a government inquiry found that architects were still specifying timbers from Baltic ports because they were familiar with their properties. Moreover, the tax differential was such that some merchants even shipped Baltic timbers across the Atlantic so that they became American for taxation purposes when shipped back to Britain (Yeomans, 1989). Legislation is not always effective for achieving the desired aims.

Sustainability is a rapidly developing concern, with a number of decision-making tools being produced to assist the designer and specifier, so that it is an issue that both design offices and individual specifiers need to be addressing. In drawing up specifications, consideration needs to be given to:

- the conservation and reuse of materials
- the conservation of energy
- the reduction of waste
- the reduction of pollution.

On an individual level, some specifiers have always pursued a policy of trying to specify products that have been manufactured locally. Their reasons are:

- They can get to site quickly if there is a problem during the building's assembly.
- It goes some way to supporting the local economy.
- It reduces transportation costs (and also pollution from unnecessary transportation).

While such a policy is admirable, it is not always feasible and the specifier has a duty to specify the product that best suits the client's particular requirements at the time.

Liability

This brings us naturally to the question of professional liability, which appears to be an important influence on the behaviour of individual specifiers and their offices. Such considerations hinder the process of awareness through the gatekeeping mechanisms that are used. Together with the pressures of time on the designer, they reinforce the tendency to adhere to the palette of favourite products. The dual considerations of liability and time are the most important factors that ensure that specifiers are not the innovators that they like to think they are. Nevertheless, it is in the interests of both building manufacturers and the publishers of architectural journals to reinforce this self image.

An obvious way to reduce liability is to use products that have some kind of guarantee. As these have little worth if the company providing the guarantee ceases trading, the tendency is to use long established companies with good track records. Another means of limiting liability is to use standard details supplied by the manufacturer. Companies who can offer specifiers these ways of limiting their risk are more likely to be able to market new products successfully. It is in a manufacturer's interest to present its product in a way that reduces both the perceived risk and the design time that the specifier will have to invest in adopting his product. Competent and readily available technical support appears to be far more effective than the employment of salespeople. However, good technical support, if not immediately available, may be useless. Again, it is a question of timing.

Small manufacturers may be unable to provide an adequate level of technical staff, but this is sometimes done through trade associations that may in effect be marketing organizations. Of course, large manufacturers also use such associations in a similar capacity, and these are often important in providing much-needed technical information. Yeomans (1988b) has noted the importance of the International Trussed Plate Association in the adoption of trussed rafter roofs; in this case, acting to prevent their rejection by the industry. However, the present research not been able to explore the extent to which information provided by such trade associations could reduce the design time and perceived risk of the product.

Office management

It is clearly in the manufacturer's interest to encourage innovative behaviour. This does not simply mean encouraging specifiers to use the manufacturer's latest product development: it means encouraging them to use any of their products that may be new to them. One may then ask whether it is in the interest of the office to encourage innovative behaviour. The answer is clearly that it is not, principally because of the penalties attached. However, there are those occasions in which the specifier has to be innovative, and the issue is how offices can be managed so that the adoption of innovations is facilitated in such circumstances.

In offices of sufficient size, or for sufficiently large projects, it may be possible to employ someone whose role is to be aware of the full range of products available and the sources of information pertaining to them, the design manager or the chief specification writer. However, this will not be possible in the majority of cases because of the small size of design organizations and, rather than looking for ways of facilitating the use of a wider range of products, offices would be do better to consider the effectiveness of their gatekeeping activities. This should not be taken to imply that they should be tightened to restrict further the flow of information into the office. Rather, they should be managed in a way that encourages the flow of information on products that satisfy the requirements suggested above, i.e. those that are accompanied by information sufficient to facilitate good detailing and the provision of adequate technical support.

Best practice

This brings us to 'best practice'. Best practice may be defined in standards and/ or research reports, although in reality, it is often deemed to be the practices that best work for a particular design office. When talking to specifiers to gather insights for this book, it quickly became apparent that individuals hold very strong views on the subject of specification, with what they considered to be best practice differing considerably from their peers, and often within the same office. Many specifiers were open and honest in their opinions and claimed that although they knew that rolling specifications from one project to the next was in principle bad practice, it worked for their organization. Such habits were regarded as widespread and effective for many design offices. What this shows is that what may be viewed to be bad practice is seen by many to be the best way of specifying for their particular circumstances, a finding that supports the authors' plea for better education and professional updating in this area. Indeed, it is hoped that the contents of this book will help students and practitioners to reconsider their own specification behaviour with a view to developing best practice.

Common problems and practical suggestions

Based on this research, it is possible to suggest a number of common problems that face specifiers and design managers, with the hope that researchers and practitioners may try to tackle them. This may involve interaction between practitioners and researchers/educators or simply the ability to recognize that (often minor) improvements in working methods could make the professional business more efficient and hence more profitable. Some of the more obvious areas to tackle include effective programming and knowledge exchange.

Effective programming

Throughout this research the lack of time allocated to the detailing and specification phases of projects was a concern to specifiers. The specifiers and design managers were quick to blame clients for not paying a high enough fee to allow adequate resourcing of projects, although it was quite noticeable that none of those interviewed felt that the lack of time had anything to do with poor estimation of design effort and thus ineffective programming of work within the office. Failure to estimate the amount of time required for specific stages of design projects has been highlighted as a problem in the literature (e.g. Emmitt, 2007). In terms of the specification process we have a catch 22 situation, because unless design and engineering offices understand the behaviour of their specifiers they cannot start to estimate accurately the amount of time required to complete this aspect of the work. Based on the work reported here, design managers may consider ways in which reliable data can be extracted from timesheets to improve their understanding of the process within their own offices and so manage the design effort more effectively.

Knowledge exchange within the office

A problem common to the offices in which the research was conducted was the apparent lack of knowledge exchange between specifiers; and this appears to be linked to the lack of time to do the job correctly. There was no evidence of formal feedback mechanisms operating in the offices observed, nor was there any evidence of benchmarking activities to assess the performance of specifiers. It was also seen that as specifiers do not always discuss issues with their colleagues, there is a danger of such searches for information being repeated, thus wasting valuable time. One way of mitigating such wasteful habits is to introduce knowledge exchange events and include through project and post project reviews. However, it is recommended that specifiers also make some attempt to look at evidence-based knowledge rather than relying entirely on experiential learning.

Justifying the approach to other team members

Some specifiers were engaged in personal battles with other project team members as they tried to retain their original (proprietary) specification decision. Clearly, different members of the team hold different values and priorities from the specifier and this is usually a positive aspect of team design. However, some attempt to justify the approach taken to specification, through discussion with other project participants, may be useful and may serve to limit the amount of time spent dealing with change requests. This could be done during formal design reviews, where information is discussed and approved before proceeding to the next stage of the project. This research did not find any evidence that this was being done; indeed, the culture within the design offices was not particularly conducive to sharing information with others, unless requested. Determining the cause of this is an area in which further research is required, although recognition of the problem may help design managers to implement appropriate protocols to improve working methods.

Future directions

Throughout this book, the specification process has been portrayed as a rich and rewarding subject. It has been argued that specification deserves greater attention from both academics and practitioners if real improvements in quality and value are to be achieved. By way of an epilogue, it is useful to look, briefly, at future developments in this area.

Specification trends

Mention has already been made of the worldwide move towards a performance-based approach to building design. Whether or not practitioners want to specify buildings and their components entirely through the use of performance specifications remains to be seen. The current argument in the research community is for their exclusive use, but practitioners appear to be happier with the proprietary method, or a mixture of performance and proprietary methods. This is almost certainly attributable to the psychology of the designer. Architects enter the profession because they are interested in architecture and the design process. Many see themselves as controlling the design and the final appearance of the building and may be loath to relinquish part of this control to others. Moreover, their skills are essentially graphic rather than literary, and developing the former rather than the latter is certainly a major feature of their education. Even so, there is still a need to raise the standard of specification writing and the decision making process that precedes it.

However, we have started to see growing emphasis on best value, and with it a move towards collaborative working. The result of this is that it is starting to become common for the design team to leave some of the specification 'open' to allow the contractor choice in the selection of building products that provide best value. Theoretically, this allows the design team to benefit from the contractor's knowledge, although it does rely on the parties to the project trusting one another and acting with integrity. Such an approach also raises issues about what constitutes best value and to whom.

Information technology and design management

Information technology (IT) is developing rapidly within construction, and various tools are available to assist the specifier with time consuming tasks. Information searches and retrieval, specification writing, and the tracking of products from manufacture to assembly on site are all made easier and more efficient through the adoption of appropriate IT. Similarly, advances in building information models (BIMs) and rapid prototyping have helped to integrate the specification process with design and realization phases. Such tools will continue to evolve, and should be welcomed. However, one should not lose sight of the fact that IT is only a tool and that it is the individuals who still need to make the decisions, confirm them, and take responsibility for the consequences of these actions.

There has also been a considerable growth of interest in practical design management techniques over the past decade, in part owing to greater attention from academics, but largely in response to an ever more competitive marketplace for professional services. Necessity has brought about greater awareness of management techniques with (mostly) improved performance as design offices learn to manage their knowledge assets creatively. Indeed, from the conversations with practitioners, it appears that placing specification in a design and information management context can help an organization to improve its effectiveness and its ability to compete. Better awareness of time constraints and careful programming can both assist the specifier and help to ensure that the office is working profitably. Moreover, the better the management control of design, the less the office will be exposed to claims.

Manufacturers' input

The trend for greater involvement and cooperation between the manufacturers of building products and those who carry out the specifying is welcome. However, many instances were observed of poor communication and a failure on the part of manufacturers to understand the nature of the design and specification process that they are trying to influence, and a need for them to have a better understanding of the behaviour and motivation of designers and specifiers through whom they have to effect that change. More effort is required on the part of many manufacturers if the communication gap between manufacture and design is to be narrowed. As well as being able to provide products of adequate quality, at a reasonable price and delivered to the site on time (the fundamentals of selling anything), manufacturers need a better understanding of the design process to market their products effectively.

Education and training

The role of education and training deserves some comment here. The authors have highlighted the fact that the specification of buildings has not been the main focus for researchers, and have raised concerns about the lack of attention given to the subject in education. To a certain extent, research into specifying will be determined by the research funding institutions' recognition that this is an important area to investigate, and hence to fund. It will also be influenced by the ability of researchers to gain access to ongoing project environments so that they are able to research and analyse current practice.

The authors believe that specification should be incorporated into design projects, with students asked to produce a written specification for a selected part of the design proposal and justify their decisions. A structured approach proposed by Abe and Starr (2003) could be utilized to great effect. This would involve students' developing an initial (outline) specification, reassessing their choice against the client brief and design parameters, and eventually finalizing their choice by writing a specification for that particular part of the design. This entire process could be documented and submitted as part of the final submission for all design projects, thus integrating the specification process with design. Supplemented with lectures, such an approach could help to equip future specifiers for a changing construction sector. Professional bodies such as the Royal Institute of British Architects (RIBA), the Chartered Institute of Architectural Technologists (CIAT) and the Royal Institution of Chartered Surveyors (RICS) also have a role to play, by including the specification of buildings within their Subject Benchmark documents and hence influencing individual educational establishments to include this within the curriculum.

Final words

On a final note, we need to remind ourselves that to specify buildings effectively and efficiently requires talented individuals who are able to synthesize and apply a wide range of knowledge in a creative manner. Specifiers, regardless of professional background, need to work in a consistent managerial framework and have access to the latest tools to help them to achieve their tasks. Properly resourced and managed, the entire decision making process known as specification is key to providing a professional service and good quality buildings that provide value to client and users alike.

References

Abe, T. and Starr, P. (2003) Teaching the writing and role of specifications via a structured teardown process, *Design Studies*, 24, 475–489.

Akin, O. (1986) *Psychology of Architectural Design*, Pion, London.

Allinson, K. (1993) *The Wild Card of Design: A Perspective on Architecture in a Project Management Environment*, Butterworth Architecture, Oxford.

AIA (1988) *AIA Handbook*, American Institute of Architects, Washington, DC.

Anderson, J., Shiers, D. and Sinclair, M. (2002) *The Green Guide to Specification: An Environmental Profiling System for Building Materials and Components* (3rd edition), Blackwell Science, Oxford.

Andrews, J. and Taylor, J. (1982) *Architecture: A Performing Art*, Lutterworth Press, Guildford.

Antoniades, A.C. (1992) *Poetics of Architecture: Theory of Design*, Van Nostrand Reinhold, New York.

Banham, R. (1969) *The Architecture of the Well-Tempered Environment*, Architectural Press, London.

Barbour Index (1993) *The Barbour Report 1993: The Changing Face of Specification in the UK Construction Industry*, Barbour Index, Windsor.

Barbour Index (1994) *The Barbour Report 1994: Contractors' Influence on Product Decisions*, Barbour Index, Windsor.

Barbour Index (1995) *The Barbour Report 1995: The Influence of Clients on Product Decisions*, Barbour Index, Windsor.

Barbour Index (1996) *The Barbour Report 1996: Communicating with Construction Customers – A Guide for Building Product Manufacturers*, Barbour Index, Windsor.

Barbour Index (1997) *The Barbour Report 1997: Electronic Delivery of Product Information*, Barbour Index, Windsor.

Barbour Index (1998) *The Barbour Report 1998: The Building Maintenance and Refurbishment Market*, Barbour Index, Windsor.

Barbour Index (1999) *The Barbour Report 1999: The Sourcing and Exchange of Information*, Barbour Index, Windsor.

Barbour Index (2000) *The Barbour Report 2000: Influencing Product Decisions*, Barbour Index, Windsor.

Barbour Index (2001) *The Barbour Report 2001: Construction Product Information: Delivery Preferences and Trends*, Barbour Index, Windsor.

Barbour Index (2002) *The Barbour Report 2002: Exploring the Web as an Information Tool*, Barbour Index, Windsor.

Barbour Index (2003) *The Barbour Report 2003: Influencing Clients: The Importance of the Client in Product Selection*, Barbour Index, Windsor.

Barbour Index (2004) *The Barbour Report 2004: Influencing Contractors – The Importance of Main and Specialist Contractors in Product Selection*, Barbour Index, Windsor.

Barbour Index (2006) *The Barbour Report 2006: Delivering Product Information Online: An Update for Building Product Manufacturers*, Barbour Index, Windsor.

Bass, F.M. (1969) A new product growth model for consumer durables, *Management Science*, 13(5), 215–227.

BEDC (1987) *Achieving Quality on Building Sites*, Building Economic Development Committee (BEDC), London.

Benes, P. (1978) The Templeton 'run' and the Pomfret 'cluster': patterns of diffusion in rural New England meetinghouse architecture, 1647–1822, *Old-Time New England*, LXVIII(3/4), Winter–Spring.

B.L. Add. MS 41133-6, British Library, London.

Blyth, A. and Worthington, J. (2001) *Managing the Brief for Better Design*, Spon Press, London.

Bowes, J.E. (1981) Japan's approach to an information society: a critical perspective, *Keio Communication Review*, 2, 39–49.

Bowley, M. (1960) *Innovations in Building Materials: An Economic Study*, Gerald Duckworth & Co., London.

Bowley, M. (1966) *The British Building Industry: Four Studies in Response and Resistance to Change*, Cambridge University Press, Cambridge.

Bowyer, J. (1985) *Practical Specification Writing: A Guide for Architects and Surveyors* (2nd edition), Hutchinson, London.

Bradbury, J.A.A. (1989) *Product Innovation: Idea to Exploitation*, John Wiley & Sons, Chichester.

Brown, L.A. (1981) *Innovation Diffusion: A New Perspective*, Methuen, London.

BS 4940 (1994) *Technical Information on Construction Products and Services*, British Standards Institution, London.

Bullivant, D. (1959) The problem of information before the architectural profession and the building industry, *Architects' Journal*, April 2, 512.

Cassell, M. (1990) *Dig it, Burn it, Sell it! The Story of Ibstock Johnsen, 1825–1990*, Pencorp Books, London.

CCPI (1987) *Code of Procedure for Production Drawings*, Coordinating Committee for Project Information, London.

CCPI (1987) *Code of Procedure for Project Specifications*, Coordinating Committee for Project Information, London.

Cecil, R. (1986) *Professional Liability* (2nd edition), Architectural Press, London.

Chermayeff, S. (1933) New materials and new methods, *Journal of the Royal Institute of British Architects*, 23 December, 165–173.

Chick, A. and Micklethwaite, P. (2004) Specifying recycled: understanding UK architects' and designers' practices and experience, *Design Studies*, 25(3), 251–273.

Chisnell, P.M. (1995) *Consumer Behaviour* (3rd edition), McGraw Hill Book Company, London.

CIAT (2006) *The Architectural Technology Careers Handbook*, CIAT, London.

Coleman, J.S. (1966) *Medical Innovation: A Diffusion Study*, Bobbs-Merrill, New York.

Concise Oxford Dictionary of Current English (1990), Clarendon Press, Oxford.

Cornes, D.L. (1983) *Design Liability in the Construction Industry*, Granada Publishing, St Albans.

Cox, P.J. (1994) *Writing Specifications for Construction*, McGraw-Hill Book Company, London.

CPIC (2003) *Production Information: A Code of Procedure for the Construction Industry*, Construction Project Information Committee, London.

Creswell, H. B. (1929). *The Honeywood File: An Adventure in Building*.

Creswell, H. B. (1930) *The Honeywood Settlement. A Continuation of the 'Honeywood File'*.

Crosbie, M. J. (1995). Why can't Jonny size a beam? *Progressive Architecture*, June 1995, 92–95.

CSI (1996) *Manual of Practice*, Construction Specifications Institute, Alexandria, VA.

Cuff, D. (1991) *Architecture: The Story of Practice*, MIT Press, Cambridge, MA.

Dallas, M.F. (2006) *Value and Risk Management: A Guide to Best Practice*, Blackwell Publishing, Oxford.

Davies, H. (2000) *The Culture of Building*, Oxford University Press, London.

Davies, S. (1979) *The Diffusion of Process Innovations*, Cambridge University Press, Cambridge.

Dixon, K. (2001) Specialist contractor procurement files. In: Emmitt, S. (ed.), *Detailing Design*, LMU, Leeds.

Donaldson, T. L. (1860) *Handbook of Specifications*, London (cited in Rosen & Regener 2005, p. 13).

Druker, P.F. (1985) *Innovation and Entepreneurship: Practice and Principles*, William Heinemann, London.

Edmonds, G. (1996) Trade literature and technical information. In: Nurcombe, V.J. (ed.), *Information Sources in Architecture and Construction* (2nd edition), Bowker Saur, London.

Egan, J. (1998) *Rethinking Construction*, DETR, London.

Egan, J. (2002) *Rethinking Construction: Accelerating Change*, Strategic Forum for Construction, London.

Ellis, R. and Cuff, D. (eds) (1989) *Architects' People*, Oxford University Press, Oxford.

Emmitt, S. (1997) *The diffusion of innovations in the building industry*, PhD thesis, University of Manchester.

Emmitt, S. (1999) *Architectural Management in Practice: A Competitive Approach*, Longman, Harlow.

Emmitt, S. (2000) 'Changing the habits of a lifetime': a critical perspective on sustainable building. In: Erkerlens, P. A., de Jonge, S. and van Vliet, A. A. (eds), *Beyond Sustainability: Balancing Between Best Practice and Utopia*, Eindhoven University of Technology, Netherlands, Keynote Paper K2.

Emmitt, S. (2001) Technological gatekeepers: the management of trade literature by design offices, *Engineering, Construction and Architectural Management*, 8(1), February 2–8.

Emmitt, S. (2002) *Architectural Technology*, Blackwell Science, Oxford.

Emmitt, S. (2006) Selection and specification of building products: implications for design managers, *Journal of Architectural Engineering & Design Management*, 2(3), 176–186.

Emmitt, S. (2007) *Design Management for Architects*, Blackwell Publishing, Oxford.

Emmitt, S. and Gorse, C. (2007) *Communication in Construction Teams*, Spon Research, Taylor & Francis, Oxford.

Emmitt, S. and Heaton, B. (2003) The introduction of Approved Document L: a study of enforced change, *Proceedings of ARCOM Nineteenth Annual Conference*, Brighton, pp. 83–89.

Emmitt, S. and Johnson, M. (2004) Observing designers: disparate values and the realization of design intent, *Building for the Future, Proceedings of CIB World Building Congress 2004*, Toronto, Paper 823 (on CD-ROM).

Foxall, G.R. (1994) Consumer initiators: both innovators and adaptors. In: Kirton, M. (ed.), *Adaptors and Innovators: Styles of Creativity and Problem Solving (revised edition)*, Routledge, New York, pp. 114–136.

Gann, D.M., Wang, Y. and Hawkins, R. (1998) Do regulations encourage innovation? – The case of energy efficiency in housing, *Building Research and Information*, 26(4), 280–296.

Gatignon, H. and Robertson, T.S. (1991) A propositional inventory for new diffusion research. In: Kassarjian, H. H. and Robertson, T.S. (eds), *Perspectives in Consumer Behaviour* (4th edition), Prentice-Hall International (UK), London, pp. 461–486.

Gelder, J. (1995) *Specifying Buildings: A Guide to Best-Practice*, NATSPEC Guide, Construction Information Systems Australia, Milsons Point, New South Wales.

Gelder, J. (2001) *Specifying Architecture: A Guide to Professional Practice*, Construction Information Systems Australia, Milsons Point, New South Wales.

Gilfillan, S. C. (1935) (1970 imprint) *The Sociology of Invention*, MIT Press, Cambridge, MA.

Gold, R. (1969) Roles in sociological field observation. In: McCall, G. and Simmons, J. (eds), *Issues in Participant Observation: A Text and Reader*, Addison Wesley, London.

Goodey, J. and Matthew, K. (1971) *Architects and Information*, Research Paper 1, University of York, Institute of Advanced Architectural Studies, York.

Grant, J. and Fox, F. (1992) Understanding the role of the designer in society, *Journal of Art and Design Education*, 11(1), 77–78.

Greenberg, B.S. (1964) Person to person communication in the diffusion of news events, *Journalism Quarterly*, 41, 489–494.

Gutman, R. (1988) *Architectural Practice: A Critical View*, Princeton Architectural Press, New York.

Hagerstrand, T. (1969) *Innovation Diffusion as a Spatial Process*, University of Chicago Press, Chicago, IL.

Hammersley, M. and Atkinson, P. (1995) *Ethnography: Principles in Practice (2nd edition)*, Routledge, London.

HAPM (1991) *Component Life Manual*, Housing Association Property Mutual.

Harris, C.M. and Dajda, R. (1996) The scale of repeat prescribing, *British Journal of General Practice*, 46, November, 649–653.

Hassinger, E. (1959) Stages in the adoption process, *Rural Sociology*, 24, 52–53.

Heath, T. (1984) *Method in Architecture*, John Wiley & Sons, Chichester.

Hodgins, E. (1946) *Mr Blandings Builds His Dreamhouse*, Simon & Schuster, New York.

Holden, R. N. (1998) Stott & Sons: Architects of the Lancashire Cotton Mill, Lancaster.

Hubbard, B., Jr (1995) *A Theory for Practice – Architecture in Three Discourses*, MIT Press, Cambridge, MA.

Hutchinson, M. (1993) The need to stick to the specification, *Architects' Journal*, 20 (October).

Hutchinson, M. (1995) Specification substitution: the new construction industry ill, *The Comparative Performance of Concrete Roof Tiles*, Redland Technologies, Surrey.

Kidder, T. (1985) *House*, Houghton Mifflin, Boston, MA.

Koebel, C.T., Papadakis, M., Hudson, E. and Cavell, M. (2004) *The Diffusion of Innovation in the Residential Building Industry*, US Department of Housing and Urban Development.

Larsen, G. D. (2005) *A polymorphic framework for understanding the diffusion of innovations*, PhD thesis, University of Reading.

Larsson, B. (1992) *Adoption av ny produktionsteknik på byggarbetsplatsen* (Adoption of new construction technology at a building site), PhD thesis, Report No. 30, Chalmers University of Technology, Sweden.

Latham, M. (1994) Constructing the Team, CM 2250, HMSO, London.

Layton, C. (1972) *Ten Innovations*, George Allen & Unwin, London.

Leatherbarrow, D. (1993) *The Roots of Architectural Invention*, Cambridge University Press, Cambridge.

Leefers, L. A. (1981) *Innovation and product diffusion in the wood-based panel industry*, PhD thesis, Michigan State University.

Leonard-Barton, D. (1995) *Wellsprings of Knowledge: Building and Sustaining the Sources of Innovation*, Harvard Business School Press, Boston, MA.

Lewin, K. (1946) Action research and minority problems, *Journal of Social Issues*, 2, 34–36.

Lewin, K. (1947) Frontiers in group dynamics II. Channels of group life; social planning and action research, *Human Relations*, 1, 5–40.

Macey, F. W. (1930) *Specifications in Detail* (4th edition, revised by Brooke, D. and Summerfield, J. W.), Technical Press, London.

Mackinder, M. (1980) *The Selection and Specification of Building Materials and Components*, Research Paper 17, University of York Institute of Advanced Architectural Studies, York.

Mackinder, M. and Marvin, H. (1982) *Design Decision Making in Architectural Practice*, Research Paper 19, University of York Institute of Advanced Architectural Studies, York.

MacLeod, M.J. (1999) Teaching prescribing to medical students, *Medicine*, 27(3), 29–30.

Mahajan, V. and Wind, Y. (eds) (1986) *Innovation Diffusion Models of New Product Acceptance*, Ballinger, Cambridge, MA.

March, J.G. (1994) *A Primer on Decision Making: How Decisions Happen*, Free Press, New York.

Marks, P.L. (1907) *The Principles of Architectural Design*, Swan Sonnenschein & Co, London.

Mercer, E. (1975) *English Vernacular Houses*, HMSO, London.

Midgley, D.F. (1977) *Innovation and New Product Marketing*, Croom Helm, London.

Moore, R. F. (1987) *Specification and Purchasing Within Traditional Contracting*, Technical Information Service 82, CIOB, Ascot.

Musmann, K. and Kennedy, W.H. (1989) *Diffusion of Innovations: A Select Bibliography*, Greenwood Press, New York.

Nason, J. and Golding, D. (1998) Approaching observation. In: Symon, G. and Cassell, C. (eds), *Qualitative Methods and Analysis in Organisational Research*, Sage, London, pp. 234–249.

Nawar, G. and Zourtos, K. (1994) And the walls came tumbling, *SPECnews*, October. (cited in Gelder, 1995, p. 129)

Newell, A. and Simon, H.A. (1972) *Human Problem Solving*, Prentice-Hall, Englewood Cliffs, NJ.

Nielson, M. and Nielson, K. (1981) *Risks and Liabilities of Specifications in Reducing Risk and Liability Through Better Specifications and Inspections*, American Society of Civil Engineers, New York.

Oostra, M. (1999) Initiatives for product development in the building industry: the architect as product innovator. In: Emmitt, S. (ed.), *The Product Champions*, LMU, Leeds, pp. 35–44.
Oxford Thesaurus (1991) Clarendon Press, Oxford.
Parker, J.E.S. (1978) *The Economics of Innovation: The National and Multinational Enterprise in Technological Change*, Longman, London.
Patterson, T.L. (1994) *Frank Lloyd Wright and the Meaning of Materials*, Van Nostrand Reinhold, New York.
Peat, M. (2007) Barriers to the uptake of sustainable construction materials-research outcome, *Architectural Technology*, 69(January/February), 25.
Peters, J.E.C. (1988) Post medieval roof trusses in some Staffordshire farm buildings, *Vernacular Architecture*, 19, 24–31.
Pool, I.de.S. (1983) Tracking the flow of information, *Science*, 221, 609–613.
Potter, N. (1989) *What is a Designer: Things. Places. Messages* (3rd edition), Hyphen Press, London.
Rand, A. (1943) *The Fountainhead*, Bobbs-Merrill, Indianapolis, IA.
RIBA (1991) *Architect's Handbook of Practice Management*, RIBA, London.
Rogers, E.M. (1962) *Diffusion of Innovations*, Free Press of Glencoe, New York.
Rogers, E.M. (1983) *Diffusion of Innovations* (3rd edition), Free Press, New York.
Rogers, E.M. (1986) *Communication Technology: The New Media in Society*, Free Press, New York.
Rogers, E.M. (1995) *Diffusion of Innovations* (4th edition), Free Press, New York.
Rogers, E.M. (2003) *Diffusion of Innovations* (5th edition), Free Press, New York.
Rogers, E.M. and Shoemaker, F.F. (1971) *Communication of Innovations: A Cross-Cultural Approach* (2nd edition), Free Press, New York.
Rosen, H.J. and Regener, J.R., Jr (2005) *Construction Specifications Writing: Principles and Procedures* (5th edition), John Wiley & Sons, Hoboken, NJ.
Rosen, M. (1991) Coming to terms with the field: understanding and doing organisational ethnography, *Journal of Management Studies*, 28(1), 1–24.
Rowe, P.G. (1987) *Design Thinking*, MIT Press, Cambridge, MA.
Sabbagh, K. (1989) *Skyscraper: The Making of a Building*, Macmillan, London.
Saint, A. (1987) *Towards a Social Architecture: The Role of School Building in Post-War England*, Yale University Press, London.
Salzman, L.F. (1952) *Building in England Down to 1540: A Documentary History*, Clarendon Press, Oxford.
Sharp, D. (1991) *The Business of Architectural Practice* (2nd edition), BSP Professional Books, Oxford.
Shoemaker, P.J. (1991) *Communication Concepts 3: Gatekeeping*, Sage, Beverley Hills, CA.
Simon, H.A. (1969) *Sciences of the Artificial*, MIT Press, Cambridge, MA.
Slaughter, E.S. (2000) Implementation of construction innovations, *Building Research and Information*, 28(1), 2–17.
Spiegel, R. and Meadows, D. (2006) *Green Building Materials: A Guide to Product Selection and Specification* (2nd edition), John Wiley & Sons, Hoboken, NJ.
Stone, P.A. (1966) *Building Economy – Design, Production and Organisation*, Pergamon Press, Oxford.
Symes, M., Eley, J. and Seidel, A.D. (1995) *Architects and Their Practices: A Changing Profession*, Butterworth Architecture, Oxford.

Tarde, G. (1903) *The Laws of Imitation* (trans. Clews Parsons, E.) (cited in Rogers, E. M., 1995), Holt, New York.

Thornley, D.G. (1963) Design method in architectural education. In: Jones, J.C. and Thornley, D. G. (eds), *Conference on Design Methods*, Pergamon, Oxford.

Utterback, J.M. (1994) *Mastering the Dynamics of Innovation*, Harvard Business School Press, Boston, MA.

Valente, T.W. (1995) *Network Models of the Diffusion of Innovations*, Hampton Press, Cresskill, NJ.

Wade, J.W. (1977) *Architecture, Problems, & Purposes: Architectural Design as a Basic Problem-Solving Process*, John Wiley & Sons, New York.

Walton Markham Associates (1981) Communicating and selling to architects, *Architects' Journal*, August, 380.

White, D.M. (1950) The 'gatekeeper': a case study in the selection of news, *Journalism Quarterly*, 27, 383–390.

WHO (1995) *Guide to Good Prescribing*, World Health Organization, Geneva.

Willis, C.J. and Willis, J.A. (1991) *Specification Writing for Architects and Surveyors* (10th edition), BSP Professional Books, Oxford.

Yeomans, D.T. (1982) Monitoring design processes. In: Evans, B., Powell, J. and Talbot, R. (eds), *Changing Design*, John Wiley & Sons, Chichester, pp. 109–124.

Yeomans, D.T. (1988a) Managing eighteenth century building, *Construction History*, 4, 3–19.

Yeomans, D.T. (1988b) The introduction of the trussed roof rafter in Britain, *Structural Safety*, 5, 149–153.

Yeomans, D.T. (1989) Structural carpentry in London building. In: Hobhouse, H. and Saunders, A. (eds), *Good and Proper Materials: The Fabric of London Since the Great Fire*, RCHM & London Topographical Society, London, pp. 38–47.

Yeomans, D.T. (1992) *The Trussed Roof: Its History and Development*, Scolar Press, Aldershot.

Yeomans, D.T. (1996) Concrete mix design: putting theory into practice. In: Emmitt, S. (ed.), *Detail Design in Architecture*, BRC, Northampton, pp. 126–135.

Yeomans, D.T. (1997) *Construction Since 1900: Materials*, Batsford, London.

Yeomans, D. T. and Smith, A. C. (2000) Alternative strategies in restoring a medieval barn. In: Kelley, S. (ed.), *Wood Structures: An East–West Forum on the Treatment, Conservation and Repair of Cultural Heritage*, ASTM STP 1351, 176–87.

Appendix: Postal questionnaire results and commentary

The results from the postal questionnaire are presented below with a commentary on the response to each question. This commentary is additional to the discussion of results presented in Chapter 8, in which the results are discussed against the Rogers model. As discussed in the main text, the postal questionnaire was useful in highlighting some of the important issues to be addressed by the observational research. The commentary is presented in italics to distinguish it from the results of the postal questionnaire.

Response

In total, 453 questionnaires were issued, of which 138 questionnaires were returned, giving a respectable response rate of 30.5 per cent.

Section 1

Job description:
Architect: 118; technician: 7; other/unknown: 13

Sample age:
Under 25: 4; 25–34: 13; 35–44: 45; 45–54: 42; 55+: 33; unknown: 1

This compares with other statistics presented by the Royal Institute of British Architects (RIBA) and in Symes et al. (1995), so the responses can be taken to be representative of a larger population of architects.

Office size (by number of technical staff recorded at the respondent's office):

	1–5 staff	**6–10 staff**	**11+ staff**
Mackinder's sample by office size	2	4	18
RIBA (1991)	70%	15%	15%
Postal questionnaire by office size	64%	17%	19%

Mackinder visited thirty six offices, of which twenty four were private architectural practices, with the remainder drawn from local government offices and architectural departments of large companies. Her sample of private architectural practices (1980: 100–101) comprised larger offices than those of the postal questionnaire respondents. The postal questionnaire respondents are close to the RIBA (1991) figures and are more representative than Mackinder's sample.

Are you responsible for running jobs?
Yes: 134 (98%); no: 4 (2%)

Average number of jobs worked on in the past 5 years:
87.5 or 17 per annum (109 responses)

The four not responsible for running jobs were those who were under 25 years old.

Project type:
Commercial: 122; residential: 115; industrial: 103; retail: 74; leisure: 73; medical: 61; other: 16

All respondents ticked at least two areas of specialization, the majority indicated three different types of project worked on in the past five years, and others indicated four or five. Thus, all respondents had claimed experience of at least two different types of project, which is consistent with other research (e.g. Symes et al., 1995).

Section 2

Q1. Listed below are some of the most popular journals: would you indicate which, in order of preference, you read?

	1st	2nd	3rd	4th	5th	6th	Total
Building Design	63	30	22	6	3	1	125
Architects' Journal	50	40	11	3	5	1	110
What's New in Building (p)	6	2	11	15	18	11	63
Building Products (p)	4	7	11	15	11	14	62
Architecture Today	5	15	16	14	4	5	59
Architectural Review	9	16	17	7	3	1	53
Building	6	10	14	8	8	3	49
Building Refurbishment	1	5	6	10	5	16	43
ABC & D (p)	2	5	2	7	10	6	32
RIBA Journal (unprompted)	10	5	4	30	1	2	25
New Builder	2	2	7	3	4	4	22
Blueprint	4	1	2	2	4	3	16
Other	3	2	4	4	–	2	15

(p): product journal.

Mackinder noted the importance of journals in staying up to date with manufacturers and products, but she was not specific about the type of journal and offered no statistical evidence. The purpose of this question, therefore, was to get an indication of the type of journal read by preference. Building Design *and the* Architects' Journal *were the most popular. The two product journals,* What's New in Building *and* Building Products *were the next most popular when all preferences were added together. However, they were mostly recorded as third, fourth, fifth or sixth choice, which suggests that they are looked at less frequently than the journals that contain more news and less product advertising.*

This is a special communication channel through which specifiers may become aware of building product innovations and part of the Rogers model. At the outset of the research it was felt that it was important to try to assess the importance of the product journals (which carry information about new building products) in relation to the professional journals (which carry some advertising). However, there is no simple, direct way of testing the extent to which products are noticed.

Q2. In your opinion, do you feel that the journals you read influence your design decisions or influence your selection of materials/products?

Yes	32	23%
Probably	73	53%
Probably not	19	14%
No	14	10%

Mackinder's sample read journals to provide a general overview of the products available. The postal questionnaire respondents clearly felt that the journals did influence their design decisions or selection of materials/products, thus emphasizing the importance of the journal (the specialist communication channel), although this was not supported by the subsequent observations.

Q3. Would you consider selecting a material/product on the strength of an advertisement or technical article in a journal?

Yes	9	7%
Probably	48	35%
Probably not	38	27%
No	43	31%

'No, but instigate checking on it'; 'Not advert alone'; 'Probably – select for further research before use'; 'No, further research needed and Agrément Certificate to be examined'; 'No, not solely'.

This question was designed to assess the function of advertisements. There was a slight tendency towards the negative, while qualifying comments confirmed that additional information was required. This tends to support Rogers' model of a search for knowledge following initial awareness.

Q4. Do you consult trade literature on a regular basis?

Yes	127	92%
No	9	7%
Unanswered	2	1%

A high number confirmed that they consulted trade literature on a regular basis, which supported Mackinder's work. However, it was not possible to ascertain whether this literature was held in a personal file of information (see Q5 below) or was from another source.

Q5. Do you keep your own file of product information?

Yes	111	80%
No	26	19%
Unanswered	1	1%

'Yes, or technical library'; 'Yes, basic information'.

A high proportion kept their own collection of literature. This supported Mackinder's findings, where architects had a 'strong tendency' to develop a file of favourite products. The response of 80 per cent was higher than the Walton Markham telephone survey, in which 60 per cent of their sample said they kept a personal file of literature. There is a problem here because the trade literature consulted (Q4) may be that kept in the respondent's own collection. The two comments received indicated that the personal collection of literature may be relatively comprehensive in that it may contain technical information from sources other than from manufacturers to a collection of 'basic information', presumably the products used most frequently. The personal collection of literature is important because it is specific to the specifier and not part of the Rogers model, but the manner in which literature gets to be included in the personal collection and the influence of it in terms of consideration of building product innovations can only be addressed by observational research.

Q6. When you receive product information from a manufacturer, do you: (a) expect to be able to make a full and detailed specification on the strength of it; (b) call a representative to assist with the specification?

	(a)	(b)			
		Yes	**Sometimes**	**Rarely**	**No**
Yes	30	5	19	4	2
Generally	57	13	30	13	1
Generally not	39	11	26	2	–
No	10	5	4	1	–
Totals (unanswered 2)		34 (25%)	79 (57%)	19 (14%)	3 (2%)

The answers to parts (a) and (b) have been combined. For example, of the respondents who expected to be able to make a full and detailed specification on the strength of information from the manufacturer (they ticked the Yes box), five called a representative to assist with the specification, nineteen called a representative sometimes, four rarely, and two did not.

(a) 'Yes, vital'; 'Expect yes, more often than not though it cannot be done', 'Usually requires verbal contact'.
(b) 'Yes, vital'; 'Yes, often have to, unfortunately; literature often inadequate'; 'Representative requested to assist if product unfamiliar, i.e. "Wonderproduct"'.

This question arose out of Mackinder's work and was designed to assess the purpose of trade literature, i.e. was it something to specify from, or more of a tool to make the specifier contact the manufacturer, thus triggering a visit from the trade representative? Some of Mackinder's sample required 'basic information', while others expected a sample specification and detailed drawings to be included in the literature. Her sample questioned the quality of the literature, which was reinforced by the comments received in the postal questionnaire. In this survey, even those who expected to be able to make a full and detailed specification on the strength of the literature also telephoned the representative to assist with the specification, which tends to support Rogers' model of the change agent as an agent of reinforcement.

Q7. When selecting materials, do you decide:

	Always	Often	Rarely	Never
Precise trade name/material	19	98	16	–
Use a generic description (unanswered 2)	3	80	38	1

Respondents could and did tick both boxes, recording a preference for using precise trade names of products over a generic description. The use of precise trade name and generic terms concurrently supported Mackinder's findings.

Section 3

Q8. Which of the following contractual arrangements have you used in the last 5 years?

	1st	2nd	3rd	4th	5th	6th	Total
Traditional contracts	127	12	–	2	–	–	141
Design and build	8	63	7	1	1	–	80
Project management	–	7	9	2	–	–	24
Management contracting	–	9	2	3	2	1	17
Construction management	1	2	4	–	2	–	9
Joint venture	–	–	5	1	1	–	7
British Property Federation	–	–	1	–	–	2	3
Other (ACA) (unanswered 4)	–	–	1	–	–	–	1

Section 3 was designed to gather information specific to the building industry and outside Rogers' general model. A shift away from traditional contracts might produce a change in specification pattern, but traditional contracts were clearly the most popular and used in preference to design and build. This was the same as Mackinder's sample, despite a difference of twelve years between the two pieces of research. Thus, comparisons between the postal questionnaire results and Mackinder's observations are valid since both are primarily concerned with traditional contracts.

Q9. Do you find that the type of contract influences the products/materials you select?

Yes	24	17%
Generally	32	23%
Generally not	40	29%
No	39	28%
Unanswered	3	3%

Yes/generally 40 per cent, generally not/no 57 per cent. There is not a sufficiently clear difference to be sure of actual behaviour here, especially since we are seeking respondents' views, and they may well wish to believe that their behaviour is unaffected.

Q10. On your last project, did you change a material/product or component during the contract as a direct result of a request by the contractor?

Yes	70	51%
No	62	45%
Unsure	6	4%

Yes comments received:

Non availability/delivery times	31
Cost	12
To suit programme	5
Design and build contract	5
Contractor's request/advice	5
To simplify construction	2

Other specific comments:

'His idea was better'. 'Cooperation'. 'Following discussions with design team and contractor'. 'Reductions in tender price offered and accepted by our QS (product similar to the one specified)'. 'Was the absolute equivalent and benefited the contractor cost wise'. 'The contractor wanted to use MDF for the window bottoms instead of the softwood specified. I had no objection'. 'Change due to new product becoming available'. 'Superior product/material'. 'The material was the equivalent of the one specified'. 'On the design and build project, the exact material is often influenced by the general contractor'. 'The alternative product proved to be equally suitable for the situation'. 'Expedience'. 'Product was equal to that originally specified'. 'At request of contractor was asked to use alternative of equal quality to that specified'. 'Often working with small contractors – wish to use materials/methods they are familiar with/easily obtainable in small quantities'.

No:

'Not requested'. (5) 'Not suitable'. 'Change nothing if possible'. 'Original spec/price had been incorporated in the bill'. 'Once specified and included in contract documents variations spell trouble'. 'The alternatives suggested by the contractor were not of suitable quality'. 'Too late to alter the design'.

Other:

'This quite often happens'.

Mackinder noted the 'widespread influence' of the contractor, although no figures were reported. This question was designed to collect some quantitative information (not available elsewhere) and qualitative information. The response is important since it confirms that the specifier's innovation decision stage is more complex than Rogers' model and supports an additional stage during which a product specified by an architect may face discontinuance through the action of the contractor.

Of the yes comments received, thirty one noted non availability/delivery times, which is primarily a problem for the contractor, not the specifier (although it may be a problem for the architect's office if the contract programme is affected). This may be the case, or it could provide evidence of an excuse to change the product to benefit the contractor. Cost was also noted, more of a concern to the contractor rather than directly to the specifier.

Some of the comments received indicated that discussion about products with people external to the architect's office took place. Some quantity surveyor (QS) involvement was also noted at this stage.

Q11. On your last five jobs, were any specific materials or components requested by any of the following?

	Often	Sometimes	Rarely	Never
Client's request	36	75	23	2
Planners' request	17	66	30	17
Contractors' request	13	64	41	8
QS suggestion	7	46	46	26
Other	2	6	2	3

This question was designed for comparison with Mackinder's sample. In her sample, both the client and the contractor were seen to be an important influence on specification decisions (the previous question confirmed the influence of the contractor). Three quarters of her sample left items to be specified by the QS, and three quarters of her sample also noted the influence of the planner. The postal questionnaire respondents recorded less influence of the QS than Mackinder's sample (this may indicate a change in practice since her work was carried out, or it may be reflective of the larger offices in her sample); otherwise, it supported her work. This is important since it is an aspect of adoption behaviour that is not comparable to Rogers' examples.

Q12. Does your practice have:

	Yes	No	Unsure
An approved list of materials/products?	43	91	3
A list of prohibited materials/products?	47	80	7

'Different list for different clients'. 'Yes, through personal experience only'. 'Unofficially a list of prohibited products'. (2)

Mackinder looked at standard specifications but did not address approved or prohibited materials. This question arose out of the group discussion and, again, is not part of the Rogers model. However, it is important since the use of such lists may prevent or encourage the use of certain products. The majority of respondents answered 'yes' to both or 'no' to both lists.

Section 4

Q13. Are you aware of price differentials between products with the same performance specification when you select them?

Always	12	8%
Often	60	44%
Occasionally	60	44%
Never	6	4%

'Quality comes first with pedigree'. 'Often – but suppliers and manufacturers often reluctant to make current prices available'.

Mackinder's sample reported cost as an important factor, but her sample complained that cost information was difficult to obtain since manufacturers did not make it available, a sentiment reported in the postal questionnaire.

Q14. Generally, when do you become aware of product cost?

	Always	Often	Occasionally	Never
Outline proposals	7	39	45	21
Scheme design	10 (17)	58 (97)	38 (83)	8 (29)
Detail design/specification	40 (57)	65 (162)	16 (99)	1 (30)
Production information	39 (96)	50 (212)	18 (117)	2 (32)
Tender stage	48 (144)	38 (250)	20 (137)	4 (36)

Cumulative totals are shown in parentheses.

'Varies depending upon size of project'. 'Ongoing general awareness with experience'.

Mackinder reported that cost awareness improved as jobs progressed and the QS became more involved, which is supported in the information reported here.

Q15. Assuming you were aware of a range of product costs, do you feel it would influence your final selection?

Always	7	5%
Often	77	56%
Occasionally	50	36%
Never	4	3%

'No two products or sub contracts are ever the same'.

The comment above helped to highlight the problems with asking such a general question and reinforced the need for some observational research. Most believe that cost has some effect, but we do not know how much. It may also depend on the type of project and/or the type of client.

Section 5

Q16. Generally, how do you learn about new building techniques/methods and new products/materials?

	Always	Often	Occasionally	Never
Trade journals	21	84	18	–
Library	8	50	46	9
Direct mail	2	55	54	7
Trade representatives	2	51	62	5
Colleagues	8	46	58	9
Exhibitions	5	23	76	17

'None of these'. 'Direct mail always in the bin'.

The questions in this section were designed specifically to address the issues coming out of an early analysis of the Rogers model. This question was designed to assess how specifiers become aware of new products/building techniques, building product innovations in Rogers' terms. It highlighted the role of trade journals (not supported by the subsequent observational research). Mackinder's sample varied in their readership of journals, but viewed them as the most important source of information about new products, supported in the information reported here. Her sample noted the importance of colleagues exchanging views in the office, but this was largely in relation to products that had failed, not a source of information about new products.

The problem with this question is that it could not address whether this was passive awareness or whether the specifier searched for information about new products: active awareness. This is Rogers' chicken and egg problem, highlighting the need for some observational research.

Q17. How frequently do you select products that you have no previous experience of?

	u25	**25–34**	**35–44**	**45–54**	**55+**	**?**	**Total**
Often	–	–	–	–	–	–	–
Occasionally	1	8	28	16	13	–	66 (48%)
Rarely	3	5	15	22	16	1	62 (45%)
Never	–	–	2	4	4	–	10 (7%)
	(4)	(13)	(45)	(42)	(33)	(1)	(138)

'Depends on definition of "no previous experience", and on type of product (e.g. wall ties no, wallpaper yes)'.

The comment received indicated the problem of asking such a question and highlighted Mackinder's observation about the relative importance of products.

Results are shown against age group for this question and questions 18, 19, 20, 22 and 23. Although Rogers' earlier work stated that adopter age influenced the adopter's level of innovativeness, this has been played down in his later work. The problem here is that such an assessment could only be made against specific products and, therefore, only the total figures are used in the main text. The comments received for these questions are more significant and highlight the need for some observational research.

Q18. How long does a product have to be on the market before you would specify it?

	u25	**25–34**	**35–44**	**45–54**	**55+**	**?**	**Total**
Over 2 years	–	4	11	20	14	1	50 (36%)
1–2 years	1	7	15	5	9	–	37 (27%)
6–12 months	3	1	8	8	5	–	25 (18%)
Under 6 months	–	1	4	1	1	–	7 (5%)
Unanswered/depends	–	–	7	8	4	–	19 (14%)
	(4)	(13)	(45)	(42)	(33)	(1)	(138)

'Depends on type of product/product performance claims'. (6) 'Hypothetical, time of no concern'. (4) 'Depends on the type of product – a brick has little to prove'. 'Depends on client'. 'Well over 2 years'. '10 years'. 'The longer the better – particularly if completely innovative'. 'When it has a BBA Certificate'. '6–12 months if it was an alternative, say, type of floor vinyl, over 2 years for major items'.

'Depends on use intended, can't be answered, tend to use well known products unless won't do job'. 'Depends on the producer'.

Time is an important factor in the Rogers model, but the comments received indicate the problem with asking such a question because it depends on the importance of the product. Furthermore, the respondents are guessing how long a product has been on the market since they do not know (without checking with the manufacturer).

Q19. In the past year, have you specified products that are new to the market or new to you?

	u25	25–34	35–44	45–54	55+	Total
Yes	3	5	23	13	8	53 (38%)
No	1	8	22	28	25	84 (61%)
Unanswered	–	–	–	1	–	1 (1%)
	(4)	(13)	(45)	(42)	(33)	(138)

Thirty two of Mackinder's sample of thirty six offices (nine tenths) preferred to use products that they had used before. Eleven (one third) of her sample reported that it was office policy to avoid the use of anything new unless completely unavoidable. The respondents to the postal questionnaire appear to be more adventurous since 38 per cent claimed to have specified products in the past year that were new to the market or new to them. Of course, Mackinder's sample may have been obliged to use new products more than they would have liked. There is a difference in the question because the question used here needed to address actual behaviour rather than preferences.

Q20. If the manufacturer you normally use for a particular application does not produce your exact requirement, do you:

	u25	25–34	35–44	45–55	55+	?	Total
Attempt to find alternative (p)	3	8	35	29	25	1	101
Find alternative/compromise	–	1	4	3	2	–	10
Ask to revise/find alternative	1	1	2	3	1	–	8
Ask to revise product (p)	–	1	1	2	2	–	6
Alternative/compromise/revise	–	2	1	3	–	–	6
Compromise and specify familiar (p)	–	–	1	1	3	–	5
Revise/compromise	–	–	–	1	–	–	1
Unanswered	–	–	1	–	–	–	1
	(4)	(13)	(45)	(42)	(33)	(1)	(138)

(p): prompted

This question was designed to assess manufacturer loyalty (which may have formed a barrier to looking for building product innovations). This was found to be low and was confirmed by subsequent observations.

Q21. The following products have been launched onto the market within the past 12 months: would you please indicate those you are aware of, those you have considered using and those actually specified?

Product	Aware	Considered	Used
Product 1	69	15	5
Product 2	66	17	3
Product 3	25	8	2
Product 4	31	3	2
Product 5	37	2	4
Product 6	19	0	0
Product 7	48	6	1
Product 8	24	13	1
Product 9	18	4	0

Twenty of the 138 respondents indicated that they were unaware of any of the products.

'I would point out that I would become aware of new products by reading trade literature or hearing about them, rarely a rep. calling to inform the office'. 'None'. (2) 'Never heard of any of them'.

This was designed as a cross check to the diary of adoption and is discussed in Chapter 9. Please note that proprietary names were used on the postal questionnaire. These have been coded products 1–9.

Q22. If you received details of an 'innovative' product, would you:

	u25	25–34	35–44	45–55	55+	?	Total
Wait until someone has specified (p)	2	5	17	22	19	–	65
Specify it on the next job (p)	–	4	12	9	2	–	27
Find out more	–	2	5	7	5	1	20
Unanswered	1	1	6	3	2	–	13
Dismiss it as too adventurous (p)	–	–	1	1	4	–	6
Wait for suitable opportunity	–	1	2	–	–	–	3
Depends	–	–	2	–	1	–	3
Invite representative	1	–	–	–	–	–	1
	(4)	(13)	(45)	(42)	(33)	(1)	(138)

(p): prompted

'Investigate technical details if interested/explore possibilities'. (16) 'Specify it on the next job if appropriate/after thorough research'. (9) 'Wait until someone else has specified it and check its performance/wait to see results'. (3) 'Hypothetical – retain for consideration'. (2) 'Investigate the producer'. (2) 'Depends on the product'. (2) 'Request test results and certificates, BBA, BSI, TRADA, etc.'. (2) 'Depends on job, use tried products unless "new" needed to do job'. 'Assess it myself, thoroughly and test it to destruction'. 'Specify it on the next job only if it was the best solution to the design problem in hand'. 'Depends if relevant – innovation does not prevent specification'.

The tendency to wait supported the views of Mackinder's sample and supported the opinions recorded in Q18. It also supports the Rogers model, where only a small percentage of a social system are classified as innovators or early adopters. The comments suggest that the product would be investigated further if of interest, again supporting the Rogers model of initial awareness leading to a search for knowledge and the start of the innovation decision process.

Q23. Have you ever specified products that you view to be novel?

	u25	**25–34**	**35–44**	**45–55**	**55+**	**?**	**Total**
Yes	1	2	15	12	11	1	42 (30%)
No	3	10	29	29	21	–	92 (67%)
Unanswered	–	1	1	1	1	–	4 (3%)
	(4)	(13)	(45)	(42)	(33)	(1)	(138)

Twenty two out of Mackinder's sample of thirty six (two thirds) classed themselves as conservative in their approach to selecting materials (see also the comment on Q19). Sixty seven per cent of the respondents to the postal questionnaire confirmed that they had never specified products that they viewed to be novel, thus supporting Mackinder's work. Care should be exercised in making a direct comparison here, but the answers to this question and those above indicate that approximately two thirds of this sample act in what Mackinder's sample described as a conservative manner when selecting materials. This trait was confirmed by subsequent observation.

A wide range of products was reported, from large items such as cladding panels to much smaller products such as ironmongery. This list is important since it indicated that different respondents viewed different products as novel (no doubt based on different experiences), thus illustrating the difficulty of trying to assess the innovativeness of the sample. Furthermore, it both relies on memory recall and is not specific to, say, the last five jobs the respondent had worked on. Thus, there is no way of telling whether the respondents only did this once, or more frequently. Once again, it helped to demonstrate the need for some observational work. In addition to this, the following general comments were received:

'Use of innovative product is often restricted, not by any doubt on performance, but by insurance companies indirectly and contractors' unfamiliarity directly'. 'Non critical items'. 'Yes, too often'.

Index